Insect Pheromone Technology: Chemistry and Applications

Insect Pheromone Technology: Chemistry and Applications

Barbara A. Leonhardt, EDITOR

U.S. Department of Agriculture

Morton Beroza, EDITOR

Silver Spring, Maryland

Based on a symposium sponsored by the Division of Pesticide Chemistry at the 182nd Meeting of the American Chemical Society, New York, New York, August 25–26, 1981.

ACS SYMPOSIUM SERIES **190**

AMERICAN CHEMICAL SOCIETY

WASHINGTON, D. C.   1982

Library of Congress Cataloging in Publication Data

Insect pheromone technology.
  (ACS symposium series, ISSN 0097–6156; 190)

  Based on the papers presented at the Symposium on
Chemistry and Applications of Insect Pheromone
Technology, sponsored by the Division of Pesticide
Chemistry, at the 182nd Meeting of the American
Chemical Society, New York, N.Y., Aug. 25–26, 1981.

  Includes bibliographies and index.

  1. Insect sex attractants—Congresses. 2. Phero-
mones—Congresses.
  I. Leonhardt, Barbara A., 1936–   . II. Beroza,
Morton, 1917–   . III. American Chemical Society.
Division of Pesticide Chemistry. IV. Symposium on
Chemistry and Applications of Insect Pheromone
Technology (1981: New York, N.Y.) V. Series.

SB933.5.I56   1982   628.9′657   82–8714
ISBN 0–8412–0724–0 AACR2 ACSMC8 190 1-260
                                      1982

ACS Symposium Series

M. Joan Comstock, *Series Editor*

FOREWORD

The ACS SYMPOSIUM SERIES was founded in 1974 to provide a medium for publishing symposia quickly in book form. The format of the Series parallels that of the continuing ADVANCES IN CHEMISTRY SERIES except that in order to save time the papers are not typeset but are reproduced as they are submitted by the authors in camera-ready form. Papers are reviewed under the supervision of the Editors with the assistance of the Series Advisory Board and are selected to maintain the integrity of the symposia; however, verbatim reproductions of previously published papers are not accepted. Both reviews and reports of research are acceptable since symposia may embrace both types of presentation.

CONTENTS

PREFACE

Those who have followed the pheromone field know that there have been great strides forward in virtually every aspect of pheromone technology, especially in the past decade. Aside from the biological aspects of the problem, which are not directly addressed here, substantial progress has been made in isolation, identification, analysis and synthesis of pheromones, and—most important and recently—in actual applications of insect pheromones to solve difficult pest-control problems.

Although much of the impetus toward pheromone use has been generated by the urgent demands from scientists and the public that pheromones be explored as a means of minimizing pesticide contamination of our food, fiber, wildlife, and the environment generally, interest has been heightened by a number of concurrent developments:

1. new and improved techniques and methods for isolating and identifying ever-decreasing amounts of pheromones;

2. actual identifications, or more complete identifications, of pheromones, especially those of important insect pests;

3. novel syntheses, including asymmetric ones, that were devised following the finding of optically active pheromones;

4. research findings and subsequently demonstrations showing that pheromones can be a safe and viable alternative to insecticides, or at least, a means of reducing insecticide use through integrated pest-management techniques;

5. and, most important, the entry into the market of commercial firms that have developed improved formulations and pheromone products and introduced new and unique equipment to make the use of pheromones practical in many instances.

Despite these advances, most pheromone workers will agree that the pheromone field is only beginning to blossom and that there are many questions requiring answers before the full rewards of this technology can be realized. To respond to some of these questions and to bring the record up to date, we invited leading scientists from the United States and elsewhere to share their latest findings, advances, and thoughts in their respective fields of pheromone technology. Thus, this volume contains their papers, which were presented at the Symposium on Chemistry and Applications of Insect Pheromone Technology at a national meeting of the American Chemical Society in 1981. Previous symposia were held in

1969 and 1975, the latter published as Symposium Series No. 23, "Pest Management with Insect Sex Attractants."

Because progress in pheromone technology has involved the coordinated effort of a wide variety of disciplines, it is anticipated that the subject matter in this volume may be useful to organic, analytical, agricultural, environmental and micro chemists, biochemists, entomologists, pest control operators, chemical manufacturers and formulators, insect physiologists, agriculture extension workers, life scientists, ecologists, university personnel (chemists, entomologists, zoologists, agronomists), state and federal government officials dealing with agriculture or the environment, and engineers designing specialized equipment. With continued cooperation of the various disciplines, we can be optimistic about pheromones becoming important pest-control tools in the future.

BARBARA A. LEONHARDT
U.S. Department of Agriculture
Beltsville, Maryland

MORTON BEROZA
Silver Spring, Maryland

March 1, 1982

Analysis of Chemical Communications Systems of Lepidoptera

J. H. TUMLINSON, R. R. HEATH, and P. E. A. TEAL

U.S. Dept. of Agriculture, Agricultural Research Service, Insect Attractants, Behavior, and Basic Biology Research Laboratory, Gainesville, FL 32604

Recent research has shown that the pheromone mediated behavior of lepidopterous insects is very complex. The chemical components of the pheromones are usually simple molecules, but complex mixtures involving permutations of geometry, functionality, and chain length are often required to elicit the complicated behavioral repertoire that eventually culminates in mating. To elucidate the chemical and behavioral aspects of this communications system, we have used a combination of methods including collection of the volatiles emitted by the female, analysis by high resolution capillary gas chromatography (GC), and the sequential and temporal analysis of the male's behavioral response to the pheromone blend and components thereof. New liquid phases and state of the art techniques have been developed for capillary GC to separate all the components of a pheromone blend. With these methods the chemical communication systems of Heliothis virescens (F.) and H. subflexa (Gn.) have been analyzed and certain aspects have been elucidated.

Numerous investigations of the pheromone communication systems of Lepidoptera have been conducted during the last two decades, probably because Lepidoptera are ubiquitous phytophagous pests and because, superficially, their pheromones and related behavior appear simple. Most of these investigations have involved the chemical identification of the pheromone or pheromone blend obtained from the females and subsequent evaluation of synthesized pheromones to determine whether or not they "work" as trap baits for males, or communication disruptants in the field. However, in the last five years there has been a growing body of evidence that lepidopteran pheromones and pheromone mediated behavior is much more complex than first believed. It is now clear that information regarding the chemical composition of the pheromone and the pheromone elicited behavior

is incomplete for most species. Thus, monitoring systems
employing pheromone baited traps do not always give consistent
results, representative of population densities, and reduction
of mating below the economic threshold is often difficult to
achieve by communication disruption with the pheromones that
have been identified for a species. The development of
effective practical insect control systems based on the use of
semiochemicals will depend on the development of a thorough
knowledge and understanding of the chemical communications
systems of these insects.

   For these reasons we decided to conduct an in-depth study of
the chemical communication systems of certain lepidopteran species. We chose to first investigate Heliothis virescens (F.)
(Lepidoptera: Noctuidae) because it is an important economic
pest and because of our previous experience with it (1). Additionally, males of this species can be mated with Heliothis
subflexa (Gn.) females to produce sterile hybrid males (2).
This phenomenon is the basis of genetic strategies for control
of H. virescens. It also provides an opportunity to study the
production and perception of pheromones by hybrids.

   Our initial goal was to accurately define the chemical composition of the pheromone produced by H. virescens females and
to analyze and describe the male behavior elicited by the pheromone and components thereof. Ultimately we plan to delineate
the chemical communication systems of both species and to analyze pheromone production and male behavior of hybrids and backcrosses. Hopefully the results of the latter part of the investigation will provide useful correlations with biochemical
genetic investigations being conducted on these hybrids by other
scientists.

   Thus far our investigations have been focused on the chromatographic analysis of pheromones produced and emitted by females and analysis of male behavior evoked by these pheromones.
The methods developed to conduct these investigations, using H.
virescens and H. subflexa as models, are presented here. These
methods are directly applicable to similar investigations of
other species.

Capillary Gas Chromatography

Although, with a few exceptions, the chemical components of
lepidopteran pheromones are simple molecules, complex mixtures
that include permutations of geometry, functionality, and chain
length are often produced and emitted by females. Analysis of
these pheromones requires a system capable of separating compounds differing in geometry and position of an olefinic bond
and resolving the mixtures produced by the females. Additionally, sensitivity sufficient to detect nanogram or smaller

quantities of these compounds is required to analyze the pheromone produced by only one or a few females. The only method
having the capabilities required for these analyses is capillary
GC with high resolution glass or fused-silica columns. This
method has the added advantage that the capillary columns can be
coupled to a mass spectrometer and a great amount of information
concerning the identity of each eluted compound can be obtained. The mass spectral data plus retention indices of a compound on two or three capillary columns that separate compounds
based on different characteristics provide complete and accurate
identification of most compounds.

The resolution (Rs) of compounds on capillary GC columns is
a function of the column efficiency (N effective), the ability
of the stationary phase to separate the compounds (separation
factor, α), and the ratio of the amount of time the compounds
spend in the stationary phase vs. the time the compounds spend
in the carrier gas phase (partition ratio, k'). Resolution of
two compounds can be defined by the equation:

$$Rs = 1/4 \frac{(\alpha-1)}{(\alpha)} \frac{(k')}{(k'+1)} N^{1/2}$$

Improvements in resolution on a capillary column after it has
been prepared can be made only by the adjustment of retention
time, which alters the partition ratio of the compounds, and by
the optimization of the carrier gas used. Increases in retention time which result in a partition ratio of greater than 5
afford very little improvement in resolution and are done at the
expense of analysis time. Similarly, the use of nitrogen instead of helium as a carrier gas results in an increase in resolution of 1.14 at the expense of doubling the analysis time.

The greatest change in resolution of components is obtained
through the use of a stationary phase that results in an increase in the separation factor (α) (see later). The amount
of resolution required is dependent on the need to analyze and
quantitate minor components that elute close to a major component of the pheromone blend. Figure 1 shows the effect that
the reduction in column efficiency (peak B), and the introduction of peak asymmetry (peak C) have on the separation of a 1%
minor component eluting before and after a major component
peak. The accurate determination of the 1% component peak which
is possible in Figure 1A is severely limited in 1B and 1C. The
detection of a 10% component (Figure 1D) is still possible with
the reduced column efficiency. A column coated with a stationary phase that improves α by ca. 0.01 as illustrated in Figure
1E is capable of providing an adequate determination of 1%
components even at reduced column efficiency.

The separations of Δ7- and Δ9-tetradecen-1-ol acetates
on 4 capillary columns coated with different stationary phases
are compared in Figure 2. The nonpolar OV-1 phase and the

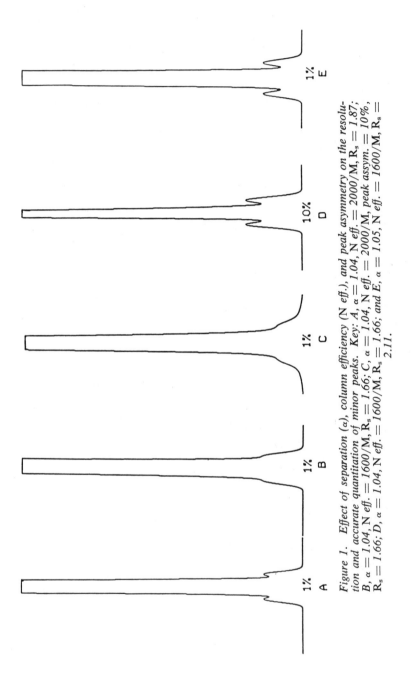

Figure 1. Effect of separation (α), column efficiency (N eff.), and peak asymmetry on the resolution and accurate quantitation of minor peaks. Key: A, α = 1.04, N eff. = 2000/M, R$_s$ = 1.87; B, α = 1.04, N eff. = 1600/M, R$_s$ = 1.66; C, α = 1.04, N eff. = 2000/M, peak assym. = 10%, R$_s$ = 1.66; D, α = 1.04, N eff. = 1600/M, R$_s$ = 1.66; and E, α = 1.05, N eff. = 1600/M, R$_s$ = 2.11.

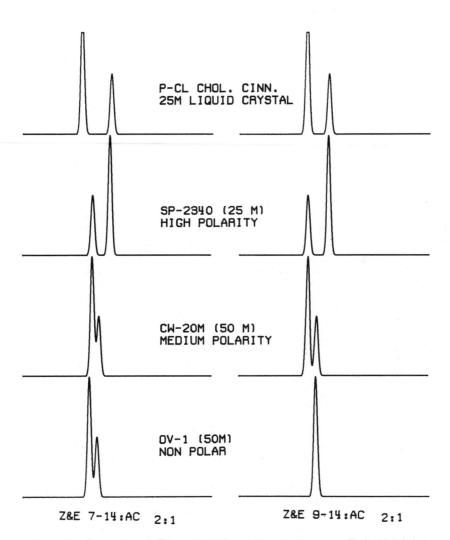

P-CL CHOL. CINN.
25M LIQUID CRYSTAL

SP-2340 (25 M)
HIGH POLARITY

CW-20M (50 M)
MEDIUM POLARITY

OV-1 (50M)
NON POLAR

Z&E 7-14:AC 2:1   Z&E 9-14:AC 2:1

Figure 2.   Separation of (Z)- and (E)-7-tetradecen-1-ol acetate (Z & E7-14:Ac) and (Z)- and (E)-9-tetradecen-1-ol acetates (Z & E9-14:Ac) on four different stationary phases on capillary columns. The OV-1 and Carbowax 20M (CW-20M) are coated on fused silica and the SP-2340 and p-chlorocholesteryl cinnamate on glass.

medium polarity Carbowax 20M are commercially available
(Hewlett-Packard) fused silica capillary columns. The usefulness of these highly efficient (N effective) fused silica
columns is severely limited because the separation factor, α,
on Carbowax 20M and OV-1 is insufficient to resolve most positional and geometrical isomers found in the pheromone blends of
lepidopteran insects. The high polarity phases containing large
amounts of cyano groups such as SP-2340 (Supelco) and Silar
10C (Applied Science) provide good resolution of many positional and geometrical isomers of mono-unsaturated compounds
found in lepidopteran insects (3). These high polarity capillary columns are commercially available. Superior resolution of
the geometrical isomers of mono and diunsaturated compounds is
obtained with liquid crystal phases, although the separation of
positional isomers is compromised as the double bond position
approaches the middle of the compound in some cases. The comparison of the separations of the analogous series of tetra-
decen-1-ol acetates on the liquid crystal and cyano phase capillary columns is shown in Figure 3. The E-isomers elute prior to
Z-isomers from the cyano phase (SP-2340). However, on the
liquid crystal column, the Z-isomers elute first when the olefinic bond is near the middle of the chain. As the double bond
is moved toward the hydrocarbon end of the chain Z and E11-14:Ac
co-elute and then the elution order reverses for Z and
E12-14:Ac. The use of both the liquid crystal columns and the
cyano columns, combined with mass spectral data on compounds
eluted from these columns, offers the most powerful analytical
procedure available for the identification of compounds like
those found in lepidopteran blends. Since use of liquid crystal
capillary columns for the separation of aliphatic olefinic
insect pheromones is a recent development, some discussion of
their properties is worthwhile.

     The use of a liquid crystal as a GC stationary phase was
first reported in 1963 (4). The application of ordered phases
to pheromone research did not occur until 1978 when Lester
reported the separation of conjugated dienes with diethyl-4,4'-
azoxydicinnamate (a smectic liquid crystal) on packed columns
(5). Subsequently we coated cholesteryl cinnamate (a cholesteric liquid crystal) on capillary columns which resulted in
coupling the separating power of the liquid crystal phases with
the high resolving capability of wall-coated open tubular
columns (6). Several liquid crystal properties must be considered when using this type of phase in GC (7). As illustrated
in Figure 4, the use of a liquid crystal such as cholesteryl
cinnamate below the temperature required for liquification of
the phase is of no utility. At its mesophase transition
temperature (temperature required for the phase to go from
crystalline to ordered liquid), which is ca. 158°C for cholesteryl cinnamate, good separation of the geometrical isomers of
tetradecen-1-ol acetate is observed. Increase in the temperature of the phase to its isotropic point (temperature at which

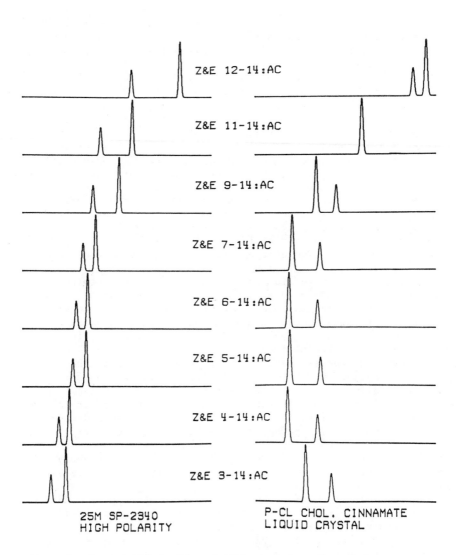

Z&E 12-14:AC

Z&E 11-14:AC

Z&E 9-14:AC

Z&E 7-14:AC

Z&E 6-14:AC

Z&E 5-14:AC

Z&E 4-14:AC

Z&E 3-14:AC

25M SP-2340
HIGH POLARITY

P-CL CHOL. CINNAMATE
LIQUID CRYSTAL

Figure 3. Separation of the (Z)- and (E)-isomers of tetradecen-1-ol acetates on SP-2340 and p-chlorocholesteryl cinnamate liquid crystal capillary columns. The ratio of Z:E is 2:1 in each set.

liquid crystal becomes unordered) results in decreased retention
time (k') observed with isotropic phases. The decrease of the
alignment of the liquid crystal molecules as the temperature
approaches the isotropic point also results in decreased α,
effective plates, and resultant resolution. If the liquid
crystal phase is first raised to its mesophase transition temperature and then gradually cooled (supercooling, Figure 4), then
decreasing the temperature below the mesophase transition point
imparts a higher degree of alignment to the liquid crystal molecules. As with an isotropic phase, retention time increases.
The more ordered phase also demonstrates increased α, effective plates, and a resultant improved resolution. A plot of
temperature of the liquid crystal vs. k', α, N/m, and resultant resolution is shown in Figure 5. The use of liquid crystal
phases at temperatures above their mesophase transition temperature dramatically deteriorates the phase's separation characteristics; however, the separations are better than those obtained
with isotropic phases like OV-1 and Carbowax 20M. Supercooling
liquid crystal phases below their mesophase transition temperatures results in increased performance at the expense of
increased retention time of the compounds. A decrease of 20°C
below the mesophase transition temperature results in a ca.
1.5-fold increase in resolution, compared with that obtained at
the crystal's mesophase temperature, which is equivalent to
increasing column length 2-fold. Although temperature considerations appear cumbersome when using liquid crystal phases, the
separations shown in Figure 6 provide the justification for
using such phases in pheromone research. The diethyl-4,4'-
azoxydicinnamate phase used in Figure 6 was first described by
Dewar (8) and recently by Lester (5) for separating pheromones
on packed columns. While this phase cannot be recommended
because it is thermally unstable and cannot be used on capillary
columns for long periods at high temperatures, it does demonstrate the potential separations possible using liquid crystal
phases. Investigations are in progress at our laboratory to
develop phases that combine the resolving power of the azoxy-
cinnamate phase with the temperature stability of cholesteryl
cinnamate phases.
    Having determined that the high polarity cyano phase and the
liquid crystal phase provide the best separation of the compounds likely to be found as components of the pheromone blend
of H. virescens and H. subflexa, we analyzed a complex mixture
of positional and geometrical isomers of 16 carbon aldehydes,
alcohols, and acetates on these two columns (Figure 7). As
noted earlier, the elution order of Z- and E-isomers is opposite
on the two phases. Aldehydes elute first on both phases. The
alcohols are retained more than the acetates on the high
polarity cyano phase, but the elution order of alcohols and acetates is reversed on the liquid crystal phase. While neither
phase separated all of the synthetic mixture, the combination of
separations obtained on both columns enabled us to pursue the

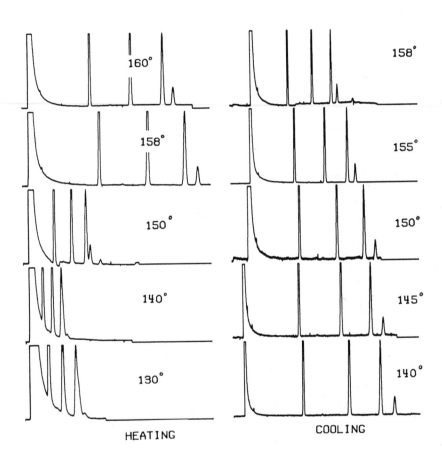

Figure 4. Effect on the separation of isomers on a liquid crystal column (20 m cholesteryl cinnamate) when increasing column temperature to mesophase transition temperature and beyond (left), and then gradually cooling (right) below mesophase transition. Peaks represent, in increasing retention time, hexadecane, heptadecane, (Z)-, and (E)-9-tetradecen-1-ol acetate.

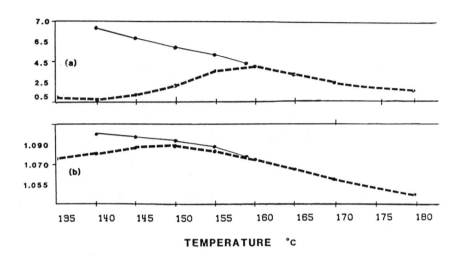

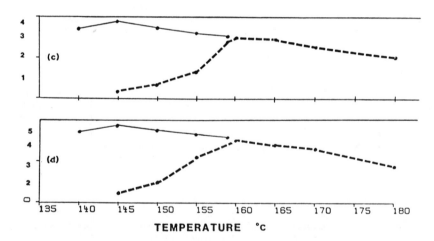

Figure 5. Effect on k' (a), α (b), N/m (1000) (c), and R (d) for (Z)- and (E)-9-tetradecen-1-ol acetate of increasing (– – –) and decreasing (——) the temperature of a cholesteryl cinnamate liquid crystal column.

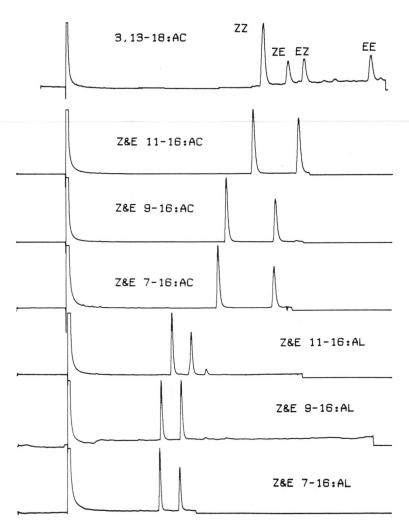

Figure 6. *Separation of various compounds on a 20 m × 0.25 mm glass capillary column coated with diethyl-4,4'-azoxydicinnamate.*

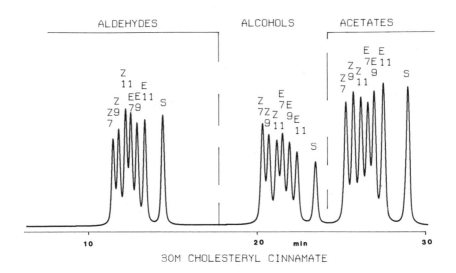

30M CHOLESTERYL CINNAMATE

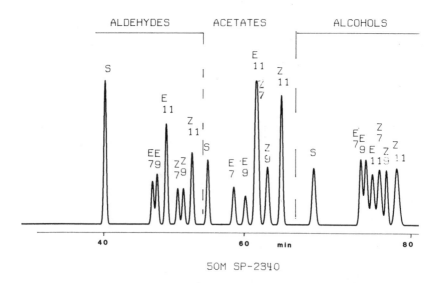

50M SP-2340

Figure 7.   Separation of a mixture of 16 carbon compounds on a liquid crystal and a cyanosilicon capillary column. S above a peak indicates the saturated compound.

identification of the components of the pheromone blend of H̲. virescens and H̲. subflexa.

Pheromone Extraction and Analysis

The most common method of obtaining the pheromone from a lepidopteran female is to clip the last few segments of the abdomen from the female's body and extract this abdominal tip with an organic solvent. Other methods include washing the body or parts of the body with solvent. Naturally, a vast number of compounds in addition to the pheromone components are extracted by these methods and must be separated. This requires a series of chromatographic purifications monitored by a bioassay to obtain the "active" compounds. The bioassay, depending on its design, may discriminate for or against certain components of the pheromone. Thus, this method often leads to the identification of only part of the components of the pheromone.

In an earlier investigation of the H̲. virescens sex pheromone we identified two compounds, (Z̲)-11-hexadecenal and (Z̲)-9-tetradecenal in a ratio of 16:1, respectively, from ether washes of calling female bodies (1̲). These compounds were also identified as H̲. virescens sex pheromones by Roelofs et al. (9̲). However, field evaluation of crude extracts and synthesized pheromone led us to speculate that other pheromone components might be present. Subsequently Klun et al. (10̲) identified seven components in the ovipositor rinses of H̲. virescens females by capillary GC. Our analysis of H̲. virescens ovipositor washes by methods similar to Klun's gave results consistent with his data. The seven components and the relative amounts found were: tetradecanal (14:Al) (1.6), (Z̲)-9-tetradecenal (Z9-14:Al) (2), hexadecanal (16:Al) (9.5), (Z̲)-7-hexadecenal (Z7-16:Al) (1), (Z̲)-9-hexadecenal (Z9-16:Al) (1.3), (Z̲)-11-hexadecenal (Z11-16:Al) (81.4), and (Z̲)-11-hexadecenol (Z11-16:OH) (3.2).

Although the female's pheromone glands would be expected to contain many compounds in addition to the pheromone, extraction of these glands, exclusive of other tissues, should provide more accurate information regarding pheromone production and composition than previous methods of extraction. Thus a histological study of the likely sites of pheromone production in terminal abdominal segments of female H̲. virescens was conducted (11̲). Two morphologically distinct areas of glandular tissue were revealed -- the most extensive situated in the intersegmental membrane (ISM) between abdominal segments 8 and 9 + 10. A second area of glandular tissue was found throughout the dorsal valves (DV). When these two glandular sites were excised separately and extracted with ether, behavioral analyses (see later) of H̲. virescens male responses indicated both glands contained active materials. However, neither extract was as effective as the whole ovipositor extract in eliciting the complete sequence of male reproductive behavior (11̲).

Capillary GC analyses on SP-2340 and OV-101 columns of the
extracts of the DV and the ISM indicated distinct differences in
the contents of these two glandular sites. The DV contained
predominantly the 14-carbon aldehydes while the ISM contained
primarily the 16-carbon compounds. Contamination of either site
by small quantities of material from the other is likely because
of the difficulty in excising the individual glands. However,
there were distinct differences in contents of pheromone components in the two sites. Although the significance of this
finding is not clear at this time, it should be noted that the
14-carbon aldehydes have not been found in other Heliothis
species and Z9-14:Al is a major factor in separating H.
virescens and H. zea (12).

The presence of a compound in a glandular extract does not
assure its emission into the air as a pheromone component.
Furthermore, the relative amounts of components emitted from the
gland may differ considerably from those contained in the
gland. There are several reports in the literature that confirm
this (13,14). Thus, to identify the pheromone accurately, we
must determine what compounds an insect emits and the ratios of
the emitted compounds. The only way to determine this is to
collect and analyze the pheromone emitted by a "calling"
female. Furthermore, we should analyze directly the collected
material since further purification may change the contents or
component ratios of the pheromone.

The method described by Grob and Zürcher (15) in which a
very small amount of charcoal is used to collect volatile compounds has been modified slightly by P. S. Beevor and coworkers,
Tropical Products Institute, London (16) to collect pheromones
from insects. We have adapted and further modified this
method. Briefly, it consists of a small charcoal filter prepared by sealing 3-5 mg of charcoal between two 325-mesh stainless steel frits in a 6 mm (O.D.), 3.7 mm (I.D.) Pyrex tube
(Figure 8). This filter is then placed at the exit end of an
aeration chamber, and air is drawn through the aeration apparatus at a flow rate of 2.5 liters/min. When aeration is complete, the filter is rinsed with six aliquots (15-20 µl) of
distilled dichloromethane; the combined aliquots are concentrated to about 5-10 µl by gently warming, and isooctane or
another solvent of choice for analysis by capillary GC with
splitless injection is added.

Evaluation of this system with standards (Table 1) indicated
that most 14- and 16-carbon aldehydes, acetates, and alcohols
could be recovered with good efficiency. Recoveries vary with
conditions, and thus it is necessary to calibrate an apparatus
with standards under the exact conditions to be used with
insects. For example, larger diameter aeration chambers reduce
the efficiency of collection because of the greater surface area
of glass available to adsorb the pheromone and lower wind velocities for a given flow rate. Therefore, the smallest aeration

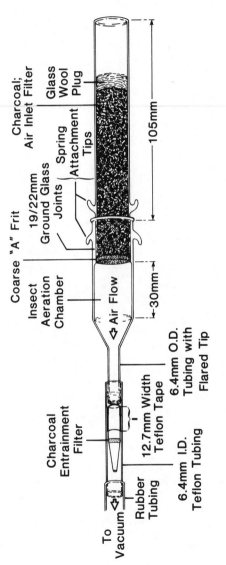

Figure 8. Volatile collection apparatus.

chamber consistent with the size and calling behavior of an
insect should be used. Nevertheless, reliable results can be
obtained when collecting pheromone with this system. Additionally, if care is taken to use pure solvents, clean glassware,
and filtered incoming air, pheromones can be collected with
minimal background impurities that interfere with analyses.

Analysis of the volatile pheromone collected from calling
female H. virescens is not yet complete. However, preliminary
results indicate that 14:Al and Z9-14:Al are emitted in greater
quantities, relative to Z11-16:Al, than are present in the
gland. Thus our data suggest that the true emitted pheromone of
H. virescens females may consist predominantly of 14:Al,
Z9-14:Al, 16:Al, and Z11-16:Al.

TABLE I

Percent of standard compounds recovered from charcoal
filter when ca. 0.5 µg of each was allowed to
evaporate from a stainless steel planchet in the
volatile collection apparatus.

Compound	% Recovered (no. replicates)
Z9-14:OH	81 $\pm$ 3 (2)
Z9,E12-14:OH	67 $\pm$ 5 (2)
Z11-16:OH	57 $\pm$ 6 (6)
Z9-14:Ac	83 $\pm$ 6 (2)
Z9,E11-14:Ac	86 $\pm$ 0 (2)
Z9,E12-14:Ac	86 $\pm$ 2 (2)
Z11-16:Ac	61 $\pm$ 10 (6)
Z9-14:Al	78 $\pm$ 6 (4)
Z11-16:Al	69 $\pm$ 9 (6)
Z11,Z13-16:Al	82 $\pm$ 3 (2)

Recently, considerable emphasis has been placed on the control of H. virescens by genetic strategies. The basis of this
approach lies in the production of sterile male hybrids from
matings between H. virescens males and females of a related species, H. subflexa (2). Subsequent matings of H. virescens males
with hybrid or backcross hybrid females produces sterile males
and fertile backcross hybrid females. Obviously, the success of
a control program based on this strategy would depend heavily on
the ability of H. virescens males to locate and mate with backcross hybrid females. Since these species are reproductively
isolated in the field despite sympatric distributions and overlapping mating periods and because interspecific matings are

difficult to obtain in the laboratory, it seemed logical that
reproductive isolation might be affected by disparate sex pheromones. Thus, we deemed it important to analyze and define the
pheromonal communication system of H. subflexa and eventually we
will analyze pheromone production and pheromone mediated
behavior of hybrids and backcrosses of these two species.

The analyses of the H. subflexa ovipositor extracts (17)
revealed eight major components coinciding in retention times on
SP-2340 and p-chloro cholesteryl cinnamate capillary columns
(Figure 9) with 16:Al, Z9-16:Al, Z11-16:Al, (Z)-7-hexadecen-1-ol
acetate, (Z7-16:Ac), (Z)-9-hexadecen-1-ol acetate (Z9-16:Ac),
(Z)-11-hexadecen-1-ol acetate, (Z11-16:Ac), Z9-16:OH, and
Z11-16:OH. Several other peaks were also variably present but,
when present, each composed less than 1% of the total mixture.
Mass spectral data confirmed the identity of the major com-
ponents.

The differences in the glandular pheromone constituents
between H. virescens and H. subflexa are distinct. H. subflexa
contains acetates of 16-carbon alcohols not found in H.
virescens and does not contain the 14-carbon aldehydes which
appear to be unique to H. virescens among Heliothis species
studied thus far. It will be interesting to see what blends the
hybrids of these species produce and relate this to data from
investigations of the genetics of this hybridization.

Behavioral Analysis

The ultimate goal of any pheromone bioassay is to define
that particular blend of pheromone components which induces the
receiver to perform a sequence of behaviors indistinguishable
from that performed during inter-organismic communication.
Typically, such analyses are conducted under laboratory conditions thereby enabling the control of such variables as light,
humidity, temperature, and insect experience, all of which have
marked effects on insect performance. However, such artificial
controls make it virtually impossible to design a laboratory
bioassay in which an insect will respond to a semiochemical
blend in the same fashion that it would in nature. Hence, the
first step in the assessment of the behavioral effects elicited
by a semiochemical blend is a critical analysis of inter-
organismic communication from which behavioral criteria can be
selected for use in further studies.

The behavioral responses of males of H. virescens and H.
subflexa to semiochemicals have been of considerable interest to
us for reasons already mentioned. Two assay systems were
employed to analyze both precourtship and close-range courtship
behaviors (18). The first, used in the analysis of male activation, orientation, and initial analysis of courtship interactions, consisted of a 1.5 x 0.5 x 0.5-m plexiglass wind tunnel
through which air was pulled at a constant rate. In our initial

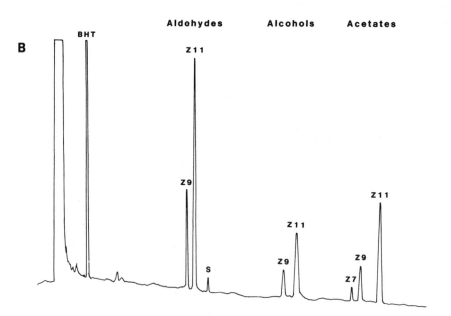

Figure 9. Analysis of H. subflexa ovipositor extract (1.5 female equivalent) on 66 m SP-2340 (A) and 30 m cholesteryl cinnamate capillary columns (B). S indicates the saturated compound.

studies, females were placed in the upwind end of the tunnel, and individual males were lowered into the airstream 10 cm from the downwind end. Male behaviors were monitored and recorded on audio cassette tapes.

In the second assay, close-range courtship interactions were observed in a 20 x 10 x 10-cm plexiglass cage. Groups of three to five females were placed in the chamber during the photophase preceding each test, and males were released individually during the scotophase when females had been observed calling for a 5-min period. Behaviors were recorded on video tape.

The behavioral responses of male H. virescens to conspecific females in these assay systems are described in detail by Teal et al. (18). Analysis of these responses allowed calculation of the statistical probability of a behavioral event occurring and enabled the selection of behaviors used in the assessment of extracts obtained from different pheromone gland sites. The sequence of behaviors performed by H. virescens males in response to the pheromone produced by calling H. virescens females and the probability of each transition occurring are given in Figure 10. This figure also shows the response of males to extracts obtained from different pheromone gland sites and to the volatiles collected from calling females.

Although the sequence of behaviors performed by males in response to calling females and the whole ovipositor extract are essentially the same, there is a much higher probability that a male will complete the sequence when responding to a female. This results from a significantly greater probability that a male will enter into taxis in response to a female, which subsequently leads to increased probabilities of performing further behaviors. Obviously, there are differences between the blends released by calling females and those extracted from the whole ovipositor (see earlier chemical analyses). Additionally, the response to either the ISM or DV is incomplete although the response to the combined ISM and DV (whole ovipositor) is complete. This is consistent with the results of the chemical analysis (see earlier) which indicated that each gland site produced only part of the total pheromone blend.

The volatile blend collected from calling females elicits responses having probabilities intermediate between those performed in response to calling females and whole ovipositor extracts. However, the volatile blend tested was not corrected for the differential recoveries of the various pheromone components (see earlier) and when corrected may increase the probabilities of behaviors occurring considerably. Nonetheless, a field trapping study using Pherocon 1C sticky traps baited with three Conrel® black fibers indicated that the volatile blend was a significantly better trap lure than were either the blend identified by Klun et al. (10) or the standard "virelure" blend (1) (Table II). Hence, this method of collecting pheromone may be the most accurate way to obtain the "true" blend emitted by the female. When the chemical analysis of this

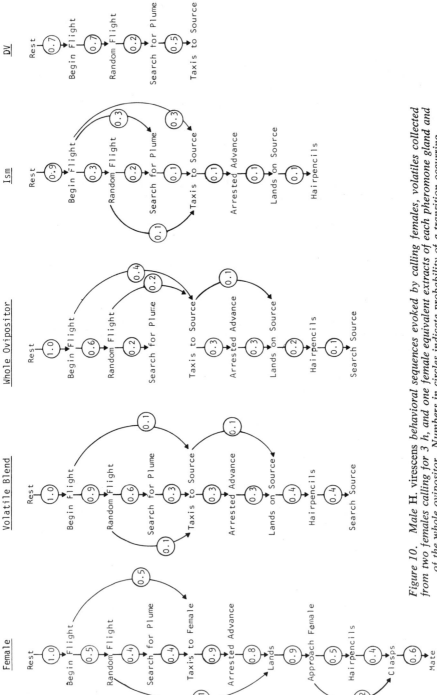

Figure 10. Male H. virescens behavioral sequences evoked by calling females, volatiles collected from two females calling for 3 h, and one female equivalent extracts of each pheromone gland and of the whole ovipositor. Numbers in circles indicate probability of a transition occurring.

material is complete, it will be interesting to observe the behavior of males to synthetic blends that correspond in composition to the various blends obtained by different methods of extraction and collection. Hopefully we can then determine the role that each component plays in eliciting the sequence of behavioral events. Additionally, we should be able to put together a synthetic blend of compounds that will be highly effective in trapping wild males in the field.

As indicated earlier H. subflexa has recently become the subject of numerous pest control studies due to the potential for population control of H. virescens by a sterile hybrid release program. However, these two species maintain reproductive isolation despite broadly overlapping ranges and intersecting reproductive periods. Hence, barriers to interspecific gene flow are most likely due to differences in their respective sex pheromone communication systems and, as indicated earlier, there are considerable differences in the pheromone gland contents of each.

TABLE 11

Trap captures of male H. virescens in sticky traps over 56 days using three Conrel fiber dispensers (ca. 8 ng/min release rate).

Synthetic blend[a]	Male trap captures[b]
Klun blend: 14:Al (1.6) + Z9-14:Al (2) + 16:Al (9.5) + Z7-16:Al (1) + Z9-16:Al (1.3) + Z11-16:Al (81.4) + Z11-16:OH (3.2) + 1% BHT[c]	10
"Virelure": Z9-14:Al (1) + Z11-16:Al (16) + 1% BHT	50
Volatile blend: 14:Al (24) + Z9-14:Al (16) + 16:Al (12) + Z11-16:Al (48) + 1% BHT	111

[a]Numbers in parentheses indicate relative amounts of each component.
[b]Mean daily trap captures were significantly different for all three blends at a 0.05 level using Duncan's Multiple Range Test.
[c]BHT = 2,6-di-t-butyl-4-methylphenol.

Therefore, field trapping tests were conducted in Florida in 1980 to study the effectiveness of synthetic blends in capturing H. subflexa males and to assess the species specificity of the pheromones of H. subflexa, H. virescens and H. zea.

The synthetic blend identical in composition to the H. subflexa gland extract (see earlier) formulated in either polyethylene vials (30 mg of mixture) or on filter paper (75 ng) was as effective in capturing males as either females or gland extracts in cone traps (19). Additionally, only H. subflexa males were captured, while traps in the same fields baited with H. virescens or H. zea females or synthetic blends (10) also captured only conspecific males.

Studies to assess the effects of deleting components of the H. subflexa blend were conducted with sticky traps which require the responding male to land to be captured. Landing is also a prerequisite for mating. The results (Table III) indicated that removal of either the aldehyde or acetate components did not completely eliminate trap capture. While the whole synthetic blend was as effective as virgin females in cone trap captures, a significant decrease in captures relative to females was noted when the whole blend was employed in sticky traps. Further, considerably more males were captured in sticky traps when the alcohols were deleted, suggesting that the alcohols may act as an inhibitor to landing. This is also supported by visual observations in the field. However, the rubber septa used as pheromone dispensers in these tests have been found to release monounsaturated alcohols at a considerably higher rate than the corresponding aldehydes. Further, subsequent testing using Conrel fibers indicated that the alcohols do not inhibit landing when dispensed in the identified ratio. Therefore, it appears that the alcohols are inhibitory only when dispensed at ratios far above those found within the gland.

Although many aspects of both the male and female reproductive behaviors of H. subflexa are similar to those of H. virescens, small differences do exist (11,18). Flight tunnel studies of the semiochemically induced behavioral interactions between H. subflexa and H. virescens indicated that males of the two species respond quite differently to the naturally released sex pheromone of the other species.

Thirteen of the 20 male H. subflexa released into the pheromone plume created by calling H. virescens females exhibited activation behaviors including wing fanning, ambulation, and genital exposure. However, only one male entered taxis up the plume and none were successful in locating females. Results of a study using one female equivalent of the H. virescens pheromone gland extract were similar, but there was a slight decrease in the number of H. subflexa males that became active and none entered taxis.

TABLE III

Comparison of sticky trap captures of H. subflexa males using different blends.[a,b,c,d,e]

Three females	16:Al	Z9-16:Al	Z11-16:Al	Z7-16:Ac	Z9-16:Ac	Z11-16:Ac	Z9-16:OH	Z11-16:OH	Mean (males/night)
					Experiment no. 1				
+									2.742 a
	+	+	+	+	+	+			1.912 a
	+	+	+	+	+	+	+	+	0.774 b
	+	+	+	+	+	+	+	+	0.706 b
						+	+	+	0.045 c
									0.000 c
					Experiment no. 2				
	+	+	+	+	+	+			2.899 (a)
	+	+	+	+	+	+			1.000 (b)
			+	+	+	+			0.941 (b)
			+	+	+	+			0.803 (b,c)
	+	+	+	+	+	+	+	+	0.578 (c,d)
	+	+	+	+	+	+	+	+	0.148 (c,d)
							+	+	0.148 (c,d)
									0.000 (d)

[a] Raw data transformed to log$_{10}$ (X+1) prior to analysis (16 replicates over 8 nights experiment 1; 10 replicates over 5 nights experiment 2).

[b] Means followed by the same letter are not significantly different in a Duncan's multiple range test at a 0.05 level.

[c] Means from the 2 experiments are not compared with one another.

[d] The presence of a compound in a test blend is indicated by a +.

[e] The relative concentrations of the compounds in each blend were identical to the relative concentrations of those particular compounds in the blend extracted from H. subflexa females.

The behavioral repertoire exhibited by male $\underline{H}$. virescens in response to the pheromone blend produced by calling $\underline{H}$. subflexa females was quite distinct from that described above. The majority (90%) of the initially inactive male $\underline{H}$. virescens flew during these tests and, in fact, the probability of undergoing taxis toward calling $\underline{H}$. subflexa females was not significantly different from that found in conspecific mating studies. However, only 75% of the responding males landed and approached $\underline{H}$. subflexa females while 93% performed these behaviors in conspecific studies. Further, the probability of interspecific mating was only 15%. Results of experiments using one FE of the $\underline{H}$. subflexa sex pheromone gland extract were considerably different from tests in which calling females were used. In fact, none of the $\underline{H}$. virescens males entered taxis up the plume. This suggests that the blend of compounds released from the extract was quite different from that released by calling females.

The blend of pheromone components identified from $\underline{H}$. subflexa female gland extracts has two components in common with $\underline{H}$. virescens, Z9-16:Al and Z11-16:Al and three not found in $\underline{H}$. virescens, (Z)-7-hexadecen-1-ol acetate (Z7-16:Ac), (Z)-9-hexadecen-1-ol acetate (Z9-16:Ac) and (Z)-11-hexadecen-1-ol acetate (Z11-16:Ac). Field studies (Table III) indicate that at the very least, the two C_{16} aldehydes plus Z11-16:Ac are of importance to the capture of male $\underline{H}$. subflexa. Hence, the absence of these three acetate components from the $\underline{H}$. virescens pheromone ($\underline{10}$), and different ratios of Z9-16:Al to Z11-16:Al probably account for the semiochemical isolation between $\underline{H}$. subflexa males and $\underline{H}$. virescens females. However, the release of pheromone components by female $\underline{H}$. virescens such as 14:Al and Z9-14:Al may be disorienting to male $\underline{H}$. subflexa and cannot be discarded as a possible mechanism for reproductive isolation.

It appears that semiochemically imparted reproductive isolation between $\underline{H}$. virescens males and $\underline{H}$. subflexa females results from the release of a pheromone blend that does not provide a stimulus of sufficient magnitude to induce a high percentage of the males to land and subsequently enter into courtship. Chemically, this may result from the distinct ratio of Z9-16:Al to Z11-16:Al produced by female $\underline{H}$. subflexa, and the probable absence of both 14:Al and Z9-14:Al from the $\underline{H}$. subflexa blend. However, because a number of males do perform courtship behaviors after contacting the female, disparities between the pheromone blends of $\underline{H}$. virescens and $\underline{H}$. subflexa are not solely responsible for reproductive isolation. Rather, it seems that both the pheromone blend and the females' ability to escape from courting males function as major inputs to reproductive isolation.

Acknowledgments

Mention of a commercial or proprietary product does not constitute an endorsement by the U. S. Dept. of Agriculture.

Literature Cited

1. Tumlinson, J. H.; Hendricks, D. E.; Mitchell, E. R.; Doolittle, R. E.; Brennen, M. M. J. Chem. Ecol. 1975, 1, 203-214.
2. Laster, M. L. Environ. Entomol. 1972, 1, 682-687.
3. Heath, R. R.; Burnsed, G. E.; Tumlinson, J. H.; and Doolittle, R. E. J. Chromatogr. 1980, 189, 199-208.
4. Kelker, H. Anal. Chem. 1963, 198, 254.
5. Lester, R. J. Chromatogr. 1978, 156, 55-62.
6. Heath, R. R.; Jordan, J. R.; Sonnet, P. E.; and Tumlinson, J. H. J. HRC & CC 1979, 2, 712-714.
7. Heath, R. R.; Jordan, J. R.; Sonnet, P. E. J. HRC & CC 1981, 4, 328-332.
8. Dewar, M. J. S.; Schroeder, J. P. J. Org. Chem. 1965, 30, 3485-90.
9. Roelofs, W. L., Hill, A. S., Carde, R. T., and Baker, T. C. Life Sciences 1974, 14, 1555-1562.
10. Klun, J. A.; Bierl-Leonhardt, B. A.; Plimmer, J. R.; Sparks, A. N.; Primiani, M.; Chapman, O. L.; Lepone, G. P.; and Lee, G. H. J. Chem. Ecol. 1980, 6, 177-183.
11. Teal, P. E. A. Ph.D. Dissertation, University of Florida, Gainesville, FL, 1981.
12. Tumlinson, J. H. "Advances in Pesticide Science" Geissbühler, H.; Brooks, G. T.; Kearney, P. C., Eds., Pergamon Press, 1979, Vol. II, p. 315-322.
13. Cross, J. H.; Byler, R. C.; Cassidy, R. F.; Silverstein, R. M.; Greenblatt, R. E.; Burkholder, W. E.; Levinson, A. R.; Levinson, H. Z. J. Chem. Ecol. 1976, 2, 457-468.
14. Hill, A. S.; Carde, R. T.; Kido, A.; Roelofs, W. L. J. Chem. Ecol. 1975, 1, 215-224.
15. Grob, K.; Zürcher, F. J. Chromatogr. 1976, 117, 285-294.
16. Beevor, P. S.; personal communication.
17. Teal, P. E. A.; Heath, R. R.; Tumlinson, J. H.; McLaughlin, J. R. J. Chem. Ecol. 1981, 7, 1011-1022.
18. Teal, P. E. A.; McLaughlin, J. R.; Tumlinson, J. H. Ann. Entomol. Soc. Am. 1981, 74, 324-330.
19. Hartstack, A. W.; Witz, J. A.; Buck, D. R. J. Econ Entomol. 1979, 72, 519-522.

RECEIVED March 1, 1982.

Some Aspects of the Synthesis of Insect Sex Pheromones

CLIVE A. HENRICK, ROBERT L. CARNEY, and RICHARD J. ANDERSON

Zoecon Corporation, Chemical Research Department, Palo Alto, CA 94304

Considerable progress has been made over the past decade in the application of insect sex pheromones to pest control programs. As the commercial applications of pheromones have expanded, the demand for larger quantities of certain of these compounds has increased, but many of the published syntheses cannot readily be carried out on a kilogram scale. We have extensively investigated alternative synthetic methods for the preparation of pheromones in high stereochemical purity. In this paper we discuss some of the practical aspects of the larger scale synthesis (100g to kilogram quantities) of insect sex pheromones with emphasis on the types of compounds that are of potential use in agriculture.

Recently several excellent reviews have appeared which describe various aspects of pheromone synthesis (1-7). Although many elegant syntheses of pheromones have been published, few of these routes are suitable for producing kilogram quantities. Efficient synthesis of these compounds depends on developing uncomplicated synthetic routes and simple purification procedures.

Purification

The chemical and stereochemical purities of synthetic pheromones are often of critical importance for their use in survey and monitoring traps (2,5). The problems associated with inhibitors requires, in general, that the synthetic products have high chemical purity. The fact that a precise mixture of geometrical isomers is usually essential if the synthetic material is to be an effective attractant in the field requires that the synthetic route give a product of reproducible and predictable stereochemical composition.

0097-6156/82/0190-0027$09.50/0

One of the simplest methods of purification involves
crystallization of synthetic intermediates (and sometimes of
the pheromone itself) at low temperature. This technique has
been extensively used in the purification of fatty acids (8,9).
Like unsaturated fatty acids, many of the related alcohols
are very insoluble in hydrocarbon solvents at low temperature.
We routinely purify intermediates and products, when applicable,
by low temperature crystallization from pentane or hexane, using
a vacuum–jacketed glass filter similar to the one described by
Schlenk (8). Quantities of material ranging from one gram to
several hundred grams can be conveniently and efficiently purified
by this method.

Another powerful purification method, applicable to larger
scale work, involves the use of urea inclusion complexes. These
crystalline clathrates, in which the "guest" molecules are en-
trapped within cavities which appear in the host helical lattice,
have been used for many years in the separation and purification
of fatty acids (9,10) and more recently in pheromone purification
(11-14). We have made extensive use of urea clathration over
the past ten years and have found the technique to be a valuable
aid in purification. In general, for straight–chain aliphatic
compounds, the urea inclusion complexes form preferentially
with the saturated analogues, followed by the E unsaturated
analogues and then by the corresponding Z isomers. For example,
a commercial sample of (Z)-9-tetradecen-1-ol containing 3% tetradecanol and 7% (E)-9-tetradecen-1-ol can be purified by selective
clathration of both tetradecanol and the (E)-olefin. The
included materials can be readily recovered by treatment with
water. Complex formation is also influenced by the skeletal
structure and by the nature of the terminal functional group.
Different polar terminal groups can substantially affect the
selectivity of clathrate formation (14).

Several chemical purification techniques can be used to
separate Z and E isomers. For example, attempted separation of
I and II (Figure 1; components of the sex pheromone of the red
bollworm moth, Diparopsis castanea) by chromatography on silica
gel-AgNO$_3$ gives only partial resolution. However, treatment
of a mixture with liquid sulfur dioxide at -20° followed by
chromatography on Florisil to remove the unreacted Z isomer
gives the 1,4-cyclo-addition product III. Thermolysis of the
sulfolene III occurs stereospecifically (15) to give the E
diene I in high yield and 99+% purity (16). The Z isomer is
obtained by treating a mixture with tetracyanoethylene
in tetrahydrofuran (THF) at room temperature which selectively
reacts with the E diene. Chromatography of the product on
Florisil enables the separation of the Z isomer II from the
Diels-Alder adduct of the E isomer I (16). We have often used
this latter technique. For example, treatment of a mixture of
E,E and Z,E conjugated diene isomers with tetracyanoethylene
selectively removes the E,E isomer as its Diels-Alder adduct (see

I

II

III

Figure 1. Separation of two components of the pheromone of the red bollworm moth.

also refs 17 and 18). Selective urea clathration can also be used
to remove the E,E isomer from a mixture of conjugated dienes or to
remove the E isomer from a mixture of conjugated enynes (e.g. 12).

Hydrogenation of Acetylenes

One very useful method of olefin synthesis involves the
introduction of an acetylene, diyne, or enyne group into a
molecule followed by semihydrogenation or reduction of the
triple bond(s). For the preparation of a Z olefin one of the
most convenient methods is the partial selective catalytic
hydrogenation of an isolated alkyne over Lindlar catalyst
[Pd-CaCO$_3$-Pb(OAc)$_2$] in the presence of synthetic quinoline (19).
Semihydrogenation is possible because the acetylene group adsorbs
very strongly to the catalyst surface. The olefin product is
protected from subsequent reaction by being rapidly displaced
from the catalyst surface by the acetylene. When most of the
latter has been consumed, further reaction of the olefin is
possible. The superiority of correctly prepared Lindlar catalyst
is due to its relative inactivity for either hydrogenation or
stereoisomerization of Z olefins. Non-polar solvents such as
pentane or hexane are preferable to alcohols for these
hydrogenations. When the reaction is carried out at room
temperature, the product contains 1.5-4% of the corresponding
E isomer. At higher temperatures, especially in alcoholic
solvents, the percentage of the E isomer can be as high as 5-10%.
We found that if the hydrogenation of the alkyne over Lindlar
catalyst (poisoned by synthetic quinoline) is carried out at lower
temperatures (-10 to -30°) in pentane, hexane, or hexane-THF,
the resulting Z olefin contains ⩽0.5% of the corresponding E
isomer (2). A similar observation has been recently reported
by Rossi and coworkers (3, 20). We have also found with Lindlar
catalyst or with quinoline-poisoned palladium-on-barium sulfate,
that the small amount of E isomer that is produced, is formed
throughout the reaction (2). The E:Z isomer ratio in the olefin
product is nearly constant during the hydrogenation but the
ratio does increase after the alkyne is consumed. It has been
previously reported, with palladium catalysts in general, that
the formation of E olefin occurs via re-adsorption and isomerization of the Z olefin and that the stereomutation of the Z olefin
occurs only in the presence of hydrogen and is not a major problem
until most of the acetylene has been consumed (19c,21). Our
observations suggest that the E olefin can also be formed during
the hydrogenation of the alkyne by either non-stereospecific
hydrogenation or,more likely,by isomerization of the newly-formed
Z olefin on the catalyst surface before the desorption takes place.
In order to minimize the isomerization to E olefin and the
over-reduction, it is advisable to use the smallest practical ratio
of catalyst to substrate. Even with the Lindlar catalyst, a high
catalyst to acetylene ratio can result in a rapid Z to E isomer-

ization of the olefin after the disappearance of the acetylene
(22a). The presence of the small amount of quinoline as a poison
is advantageous for selectivity but the success of the Lindlar
catalyst is even more dependent on the poisoning by lead acetate.
Use of palladium-on-calcium carbonate, even in the presence of
synthetic quinoline,usually gives very poor results (e.g. ref 23).
Reproducibility of catalyst preparation can be a problem. We have
found that the activity and stereoselectivity of the Lindlar
catalyst can vary even among batches purchased from the same
manufacturer. Thus, it is best to test each batch of catalyst
for both activity and selectivity prior to its use. Also it is
usually necessary to begin with pure alkyne to prevent total
poisoning of the catalyst.

Alternative catalysts such as palladium-on-barium sulfate
(poisoned by synthetic quinoline) (24), "P-2" nickel boride
(with ethylenediamine) (25), and other nickel catalysts (19c)
can be used in place of Lindlar catalyst. However, in our hands
selective hydrogenation of triple bonds to Z olefins proceeds
with the greatest stereoselectivity with Lindlar catalyst.
Palladium-on-barium sulfate (in ethanol with quinoline) can give
considerable over-reduction and isomerization to the E isomer
(22a). Use of "P-2" nickel boride as the catalyst at room
temperature usually gives ca. 2% of the E isomer (e.g. 23).
In contrast to Lindlar catalyst we have found that the hydrogenation of an alkyne over ethylenediamine-poisoned "P-2" nickel
boride or quinoline-poisoned palladium-on-barium sulfate always
gives a minor amount of the saturated hydrocarbon in addition
to the olefin. The ratio of saturated hydrocarbon to olefin
(about 0.01) also is nearly constant throughout the hydrogenation
until the alkyne is consumed, and then it increases. Further
reaction of the alkene on the catalyst surface before desorption
would explain these results.

Partial hydrogenation of skipped 1,4-diynes or 1,4-enynes
over Lindlar catalyst proceeds stereoselectively in high yield
(e.g. 26), but the partial hydrogenation of triple bonds which
are conjugated with double or triple bonds places high demands
on the catalyst. With these compounds one does not, in general,
observe a prominent break in the rate of hydrogen uptake at the
completion of semihydrogenation. Mixtures are usually obtained
containing ≤40% of the desired conjugated diene (2,27-29).

An alternative method to the selective semihydrogenation
of an internal alkyne involves hydroboration with a sterically
hindered reagent (e.g. disiamylborane), and protonolysis of the
vinylborane intermediate with a carboxylic acid to give the Z
alkene (30). This method usually gives good yields of olefins
(≥98% Z isomer) and can even be applied successfully to the
reduction of polymer-bound alkynes to polymer-bound Z alkenes
(31). For non-conjugated alkynes we prefer to use hydrogenation
over Lindlar catalyst at low temperature because of its greater
stereoselectivity and convenience but for conjugated diynes and
enynes this hydroboration-protonolysis method is vastly superior.

Reduction of Acetylenes

The preparation of pure isolated E olefins is readily accomplished by the reduction of an alkyne with metallic sodium or lithium in liquid ammonia (27,32). This reaction is preferably carried out by the addition of the alkyne in an ether to a mixture of sodium (or lithium) in liquid ammonia at -30°. The alternative reverse addition procedure can give incomplete reduction of the alkyne (33). An increase in the ratio of liquid ammonia to alkyne (34), the addition of co-solvents (23), the use of lithium rather than sodium, or the use of a higher temperature in an autoclave are advisable for the reduction of high molecular weight alkynes to overcome solubility problems which can also result in incomplete reduction. The resulting olefin is usually very pure E isomer containing no detectable Z isomer. Use of an alcohol as a co-solvent and proton donor can accelerate the reduction, but the resulting olefin then contains a minor amount of the Z isomer. Polymer-bound alkynes can not be successfully reduced with sodium in liquid ammonia (35).

Conjugated E,E dienes cannot be prepared, in general, by the reduction of conjugated diynes with sodium or lithium in liquid ammonia (36). Also the reduction of 1,4-diynes (in the absence of added alcohol) is often accompanied by, or preceded by, base-catalyzed isomerization to a compound with conjugated unsaturation, which is then reduced to the monoene level (27). However, one can successfully reduce a 1,4-diyne or an (E)-1,4-enyne in good yield to the corresponding (E,E)-1,4-diene with sodium (or lithium) in liquid ammonia-THF in the presence of tert-butyl alcohol (2,37). Reduction of a 3-alkyn-1-ol such as IVa (Figure 2) with sodium in liquid ammonia-THF is not completely stereoselective and the product is reported to contain 2-4% of the 3Z isomer (23). However, reduction of the corresponding tetrahydropyranyl ether IVb proceeds, in our hands, with high stereoselectivity to the 3E,13Z isomer Vb. (cf. ref 23). (E)-Olefins can also be prepared stereoselectively in high yield by the slow reaction of the corresponding isolated alkyne with a large excess of lithium aluminum hydride (LAH) in diglyme at 140° (20b,38).

Reduction of an alkyne with MgH_2-CuI or MgH_2-CuOtert-Bu in THF at -78° is reported to give the (Z)-alkene in good yield with very high stereoselectivity (39). Reduction of the triple bond in a conjugated enyne with a large excess of activated powdered zinc in aqueous n-propanol proceeds with Z stereoselectivity (28,40). We have found that (E)-11-tetradecen-9-yn-1-ol is reduced under these conditions to (9Z,11E)-9,11-tetradecadien-1-ol with less than 0.1% contamination by other isomers. However, this method appears suited only to small scale work because of the large excess of zinc required.

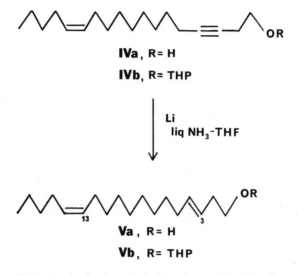

IVa, R = H

IVb, R = THP

Li
liq NH₃-THF

Va, R = H

Vb, R = THP

Figure 2. Reduction of a 3-alkyn-1-ol and its THP ether.

(9Z,12E)-9,12-Tetradecadien-1-yl Acetate (XIV)

The compound XIV occurs widely as a component of lepidopterous sex pheromones, especially of pests of stored grain and dried fruit (2). The synthesis discussed below is an example of the preparation of a "skipped" or "methylene-interrupted" 1,4-diene. It is important to have pure XIV for pheromone studies since the other stereoisomers can act as inhibitors of the behavioral response in some species.

Most of the published syntheses of XIV have been previously summarized (2). One of the more practical large-scale syntheses of XIV in high purity is the one outlined in Figures 3 and 4, which is based on an earlier synthesis by Jacobson and co-workers (41; see also 20b). Thus the diol is converted to the bromohydrin VI by reaction with aqueous HBr in a two-phase system. The crude product is purified by conversion of the alcohol group to a non-volatile borate ester with triethyl borate and removal of the 1,8-dibromooctane by distillation. After hydrolysis of the borate any residual diol is removed by extraction with water. The bromohydrin is then quantitatively converted to the THP ether, which is added to LiC≡CH-ethylenediamine complex in dimethyl sulfoxide to give VII in very high yield. No purification is required in either of these steps. A shorter synthesis of 9-decyn-1-ol (IX) was subsequently developed. An acetylene "zipper" isomerization (42) with sodio-1,3-diaminopropane is used to convert either 3-decyn-1-ol (VIII) or 2-decyn-1-ol into the terminal alkynol IX. Both of these starting alkynols are commercially available and the saving in labor is advantageous for moderate scale production.

One useful method for preparing (E)-1,4-enynes involves the cuprous salt-catalyzed coupling of 1-alkynylmagnesium halides with allylic halides in ether or THF (43). We utilized (Figure 4) the coupling of 1-chloro-2-butene in THF in the presence of dilithium tetrachlorocuprate, with the Grignard reagent prepared from VII (with EtMgCl), which gives the (E)-1,4-enyne alcohol XII, mp. 19°, in 65% yield after hydrolysis of the protecting group and purification (see below). The crude product of the coupling reaction consists of a mixture of the allylic isomers X, XI and XII in the ratio of 7.5:7.5:85, respectively, under these conditions. The isomeric composition of the product appears to be similar irrespective of the isomeric composition of the starting allylic chloride (cf. 44, 45). Thus technical crotyl chloride from Aldrich Chemical Company [containing 59% (E)-1-chloro-2-butene, 11% of the (Z)-isomer and 30% of 3-chloro-1-butene] is used without purification. However, the ratio of isomers is influenced somewhat by the reaction conditions. The straight chain isomers are favored over X by a lower reaction temperature, the use of chloride rather than bromide as the counter ion, and dilution of the reaction mixture with THF. However, practical scale-up considerations require that the reaction be run at higher con-

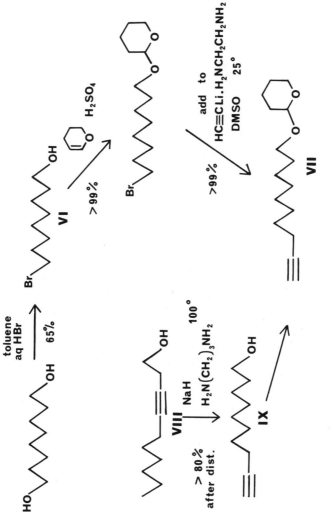

Figure 3.   Synthesis of (9Z,12E)-9,12-tetradecadien-1-yl acetate, Part 1. (see Figure 4)

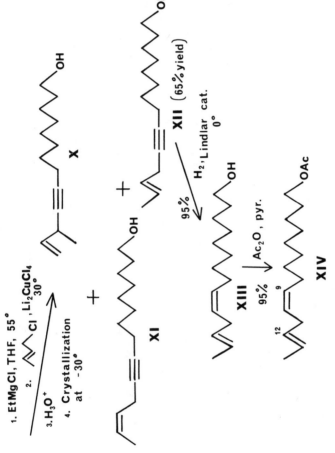

*Figure 4.   Synthesis of (9Z,12E)-9,12-tetradecadien-1-yl acetate (XIV), Part 2.
(see Figure 3)*

centration than the optimum required for regioselectivity. The
crude enyne XII is purified by filtration through an equal weight
of silica gel in ether-hexane (to remove catalyst poisons which
interfere with the hydrogenation step) followed by repeated
crystallization from hexane at -30° to remove the branched chain
isomer X and most of the Z isomer XI. The enynol XII is then
semihydrogenated over Lindlar catalyst in hexane-THF at 0° in the
presence of quinoline. Crystallization of the product from hexane
at -60° and esterification give XIV in high purity (98.4% Z,E,
1.5% Z,Z, 0.1% E,E, and <0.1% E,Z) [this 1,4-diene autooxidizes
rapidly in air to give unstable 1,3-diene hydroperoxides (46);
antioxidants such as BHT or N,N'-dioctyl-1,4-phenylenediamine
should be added to suppress this reaction (2)].

California red scale

The California red scale, Aonidiella aurantii is a worldwide pest of citrus trees and the sex pheromone produced by the
females has been identified as a mixture of the norsesquiterpenoid
esters, 3-methyl-6-isopropenyl-9-decen-1-yl acetate (XV) and
(Z)-3-methyl-6-isopropenyl-3,9-decadien-1-yl acetate (XVI)
(47,48). We have prepared the optical isomers of the former (49)
and the geometric and optical isomers of the latter (48) and
biological evaluations have demonstrated that one isomer of each
component is significantly more active than the others (50). Thus
the natural sex pheromone probably consists of a mixture of
(3S,6R)-XV and (3Z,6R)-XVI (Figure 5). Field tests have also
shown that the compounds XV and XVI are independently attractive
to males, and that there is no synergistic effect when XV and
XVI are combined. In addition, the presence of the inactive
stereoisomers does not inhibit the trap catch of males. Thus
synthetic compound for use in monitoring traps in the field can be
either XV or XVI and need not be stereochemically pure.

Since initially neither the stereochemistry of the trisubstituted double bond nor the absolute configuration of component
XVI was known, we prepared all four of the possible geometric and
optical isomers starting from either (S)-(+)-carvone (Figure 6)
or (R)-(-)-carvone, and only the 3Z,6R isomer was found to be
attractive to the males (48). This is not a practical route to
XVI; a shorter synthesis of racemic XVI was subsequently published
by Still and Mitra (51).

For field applications, we normally use component XV as a
mixture with its biologically less active stereoisomers. Our most
practical route (Figure 7) to material suitable for commercial use
produces a mixture which is only partially enriched in the S
isomer at C-3 (52; see also 53). Commercially available (S)-(-)-
citronellol of moderate optical purity (ca. 75% enantiomeric
excess; 54) is used as the starting material. The copper(I) cat-
alyzed addition of 3-butenylmagnesium chloride to 6,7-epoxy-

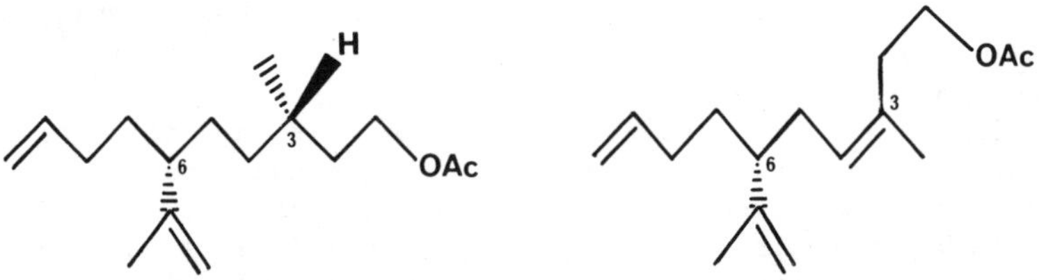

Figure 5.   Components of the pheromone of the California red scale.

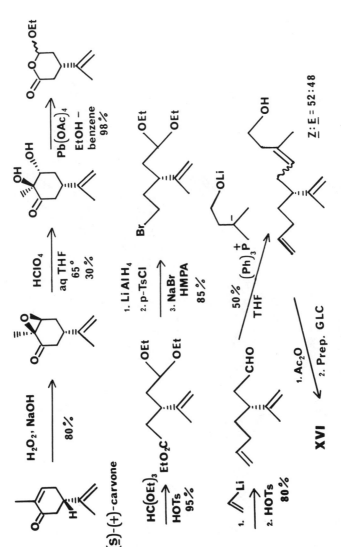

Figure 6. Synthesis of component XVI.

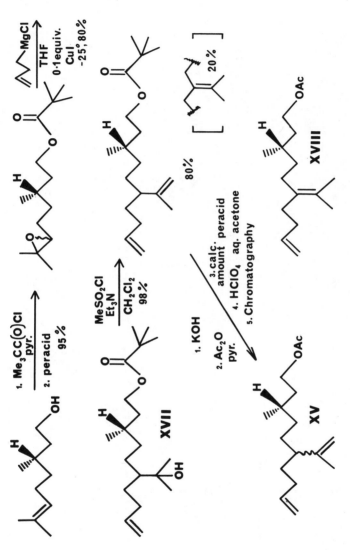

Figure 7. Synthesis of component XV.

citronellyl 2,2-dimethylpropanoate gives, in 80% yield, the
hydroxy ester diastereomers XVII, which are then converted to XV
in 60% yield as shown in Figure 7. We have also found that it is
unnecessary to remove the inactive tetrasubstituted olefinic
impurity XVIII (present in ca. 20%) from the final product for
it to be equally effective in field tests (50).

San Jose scale

The San Jose scale, Quadraspidiotus perniciosus, is a serious
worldwide pest of deciduous fruit trees. The pheromone has
been shown to be a three-component mixture of 7-methyl-3-
methylene-7-octen-1-yl propanoate (XIX), (Z)-3,7-dimethyl-2,7-
octadien-1-yl propanoate (XX), and the corresponding (E)-isomer
XXI (55, 56, 57) (Figure 8). Greenhouse and field bioassays have
shown that XIX, XX, and XXI are independently attractive to the
male San Jose scale and that the three components show almost
equal attractancy (55,57). Interestingly, neryl and geranyl
propanoate are inactive.

We have prepared the components XIX and XX by several routes
(56,57). Our first synthesis (Figure 9) utilized the addition
of a common organocopper reagent to alkyne substrates to prepare
both components (56). A better route to XX is shown in Figure 10.
The dianion of propynoic acid is alkylated with XXII and the
reaction mixture is quenched with methyl iodide to give the methyl
alkynoate. Conjugate addition of lithium dimethylcuprate then
gives XXIII, which is converted to XX as before. The yield from
this route is higher and the procedure is simpler.

A more direct approach to XIX that can be readily scaled-up
is outlined in Figure 11 (57). Reaction of the dianion of
3-methyl-3-buten-1-ol with the bromide XXIV gives the alcohol
XXV in one step. Subsequent esterification and purification
gives XIX in 40% yield from 3-methyl-3-buten-1-ol [the synthesis
of the third component XXI is described by Anderson et. al. (57)].

1,4-Disubstituted 1,3-Dienes

Conjugated dienes have been identified as components of the
sex pheromones of many lepidopterous species (2-7). Here
we would like to make a few comments on some of the better
synthetic methods for their preparation.

(Z,Z)-1,3-Dienes. The (Z,Z)-isomers can be conveniently
prepared in high stereochemical purity either from dihydroboration
of the corresponding conjugated diyne with dicyclohexylborane
followed by protonolysis of the intermediate organoborane or from
hydroboration (with disiamylborane)-protonolysis of the corresponding (Z)-enyne (60). As discussed above, the partial hydrogenation of such conjugated diynes and enynes over Lindlar catalyst is a much inferior alternative that usually gives mixtures
containing $\leqslant$40% of the desired (Z,Z)-1,3-diene. The required

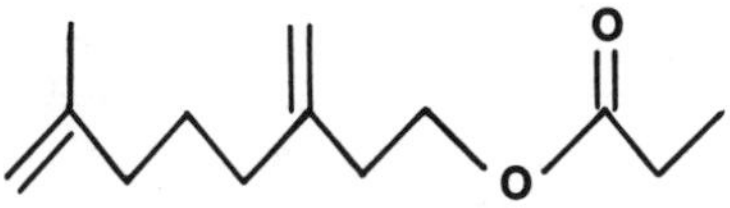

XIX

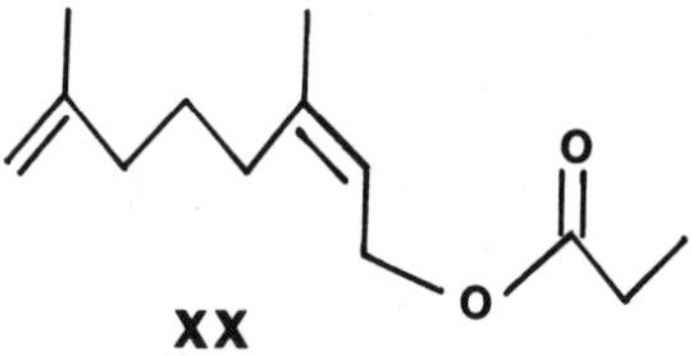

XX

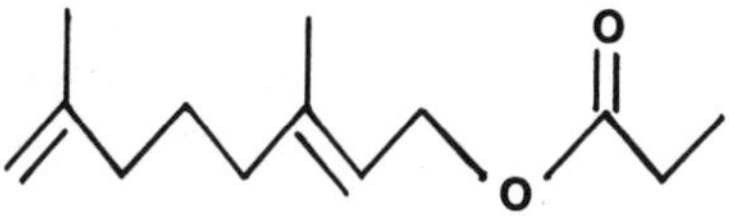

XXI

Figure 8.   Components of the pheromone of the San Jose scale.

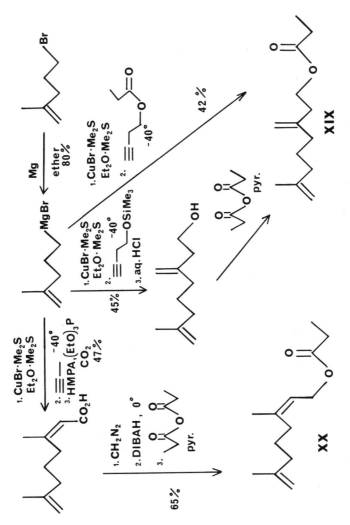

Figure 9. Synthesis of components XIX and XX.

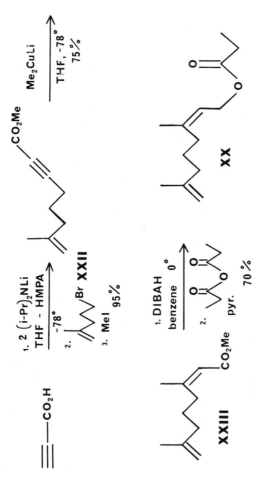

Figure 10. Improved synthesis of component XX.

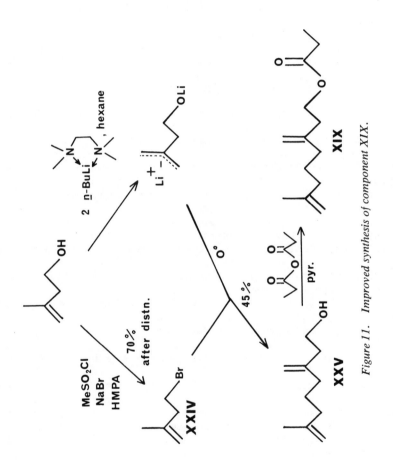

Figure 11. Improved synthesis of component XIX.

conjugated unsymmetrical diyne is usually prepared by the Cadiot-
Chodkiewicz coupling reaction between the appropriately substituted 1-alkyne and 1-bromo-1-alkyne in the presence of catalytic amounts of cuprous salts and amines (61). Two satisfactory
alternative synthetic routes to unsymmetrical 1,3-diynes, one
proceeding through dicyclohexylmethylthioborane (62) and the other
through disiamylmethoxyborane (63), are available.

Figure 12 shows a typical preparation from our laboratory
of a (Z,Z)-1,3-diene via the conjugated diyne. Figure 13 outlines
one of our syntheses (64) of (11Z,13Z)-11,13-hexadecadien-1-al
(XXVI), a component of the sex pheromone of the navel orangeworm,
Amyelois transitella (65), which proceeds through the (Z)-enyne.
The compound XXVI has also recently been prepared in high
stereochemical purity via the conjugated diyne by us (Figure 14)
and by Sonnet and Heath (66).

(E,Z)-1,3-Dienes. Several good stereoselective methods are
now available for the synthesis of unsymmetrically substituted
conjugated (E,Z)-dienes and examples of these have been recently
discussed in detail (2,6,7,67). Several of the more practical
routes involve the preparation of an (E)-1,3-enyne intermediate
followed by hydroboration-protonolysis. For example, Figures 15
(68) and 16 (69) outline two of our routes to (7E,9Z)-7,9-dodeca-
dien-1-yl acetate (XXIX), the sex pheromone of the European grape-
vine moth, Lobesia botrana (70). Both of these routes depend on
the synthesis of the key intermediate (E)-7-dodecen-9-yn-1-ol
(XXVIII) (m.p. ca. -10° to -15°) which can be readily purified
to >99.5% purity by crystallization from pentane or hexane
at -35°. Selective hydroboration of the triple bond of the
(E)-enyne with 1 equiv of disiamylborane followed by protonolysis
of the vinylborane intermediate with acetic acid gives the
required 7E,9Z diene in high stereochemical purity.

The stereoselective procedure outlined in Figure 15 is based
on the work of Negishi et. al. (71). The overall yield (from
XXVII) of pure XXVIII was only ca. 20% and although the conversion of XXVII to the THP of XXVIII is carried out in one pot, the
reaction is difficult to follow and is not practical on a large
scale. Negishi and Abramovitch (72) have prepared XXIX in
considerably higher yield by a modification of this method but in
our hands the much less stereoselective route outlined in
Figure 16 is preferable for large scale synthesis. The Claisen
rearrangement reaction of 1-hepten-4-yn-3-ol (XXX) with triethyl
orthoacetate is not completely stereoselective and gives XXXI as
a mixture of E:Z isomers in the ratio of ca. 4:1. It is not
necessary to separate isomers at this stage since crystallization
of the crude alcohol XXVIII from pentane at -35° readily affords
material of 99+% purity in >60% overall yield from XXXII.

Recently a new stereoselective method for the synthesis of
(E,Z)-1,3-dienes has been developed by Zweifel and Backlund (73)
which proceeds through a lithium dicyclohexyl[(E)-1-alkenyl]-

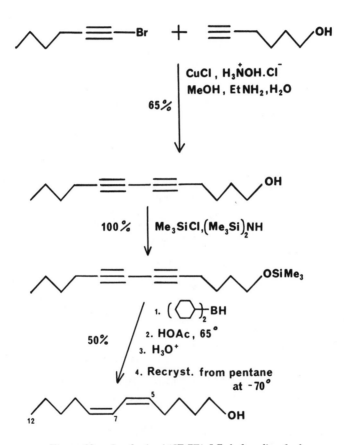

Figure 12. Synthesis of (5Z,7Z)-5,7-dodecadien-1-ol.

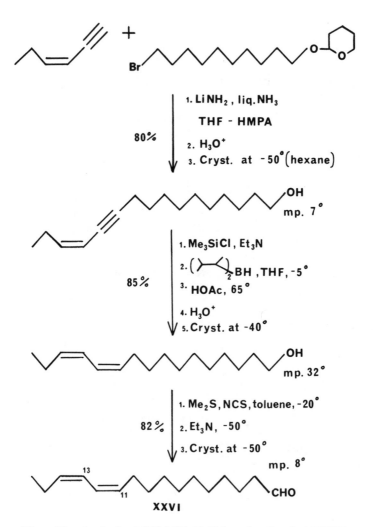

Figure 13.   Synthesis of (11Z,13Z)-11,13-hexadecadien-1-al (XXVI).

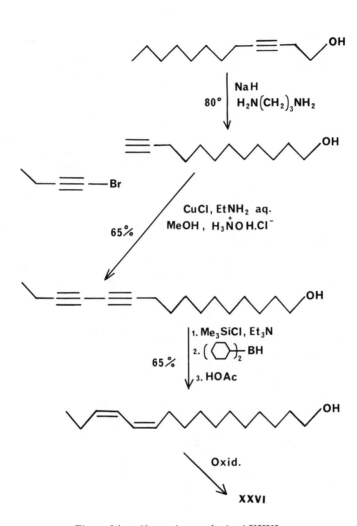

Figure 14. Alternative synthesis of XXVI.

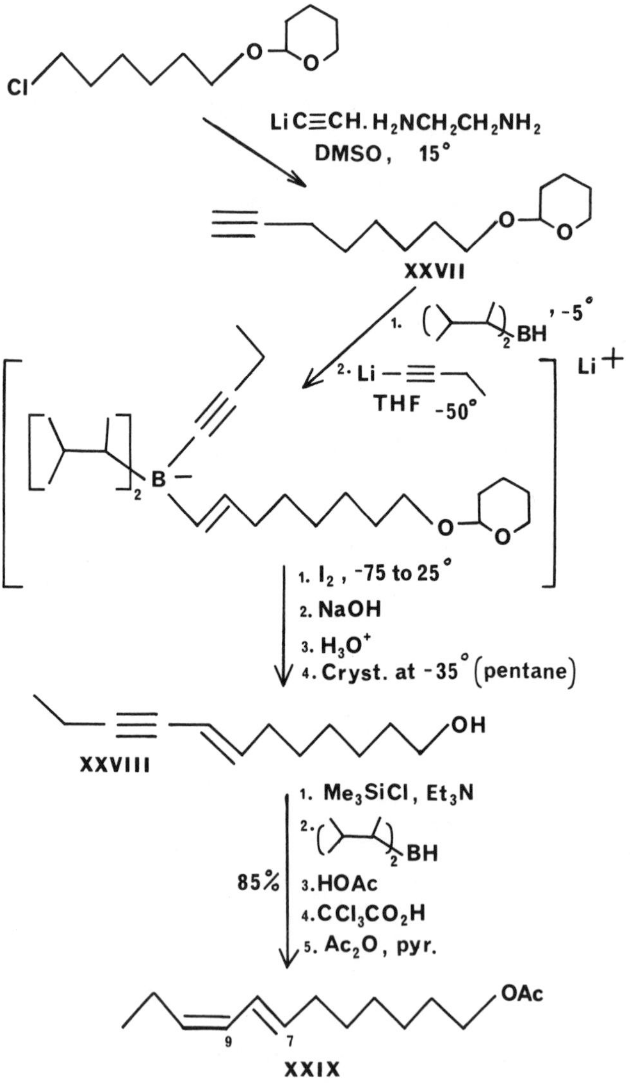

Figure 15.   *Synthesis of (7E,9Z)-7,9-dodecadien-1-yl acetate (XXIX).*

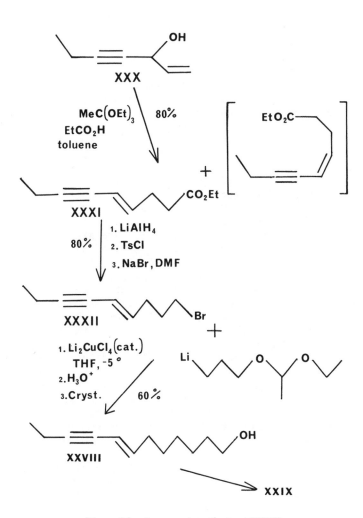

Figure 16.   *Improved synthesis of XXIX.*

(1-alkynyl)borate as the intermediate. This method affords
the diene directly without the formation of an enyne and hence
eliminates the hydroboration-protonolysis step in Figure 15.
This method is very useful for synthesis of such compounds
on a moderate scale. We have applied it to the preparation
of a number of (Z,E)- and (E,Z)-1,3-dienes including the
pheromone of Lobesia botrana, (XXIX).

For example, in Figure 17 this route is applied to the
preparation of (10E,12Z)-10,12-tetradecadien-1-yl acetate
(XXXVI). The conversion of the terminal acetylene XXXIII to the
(E,Z)-dienol is carried out in 'one pot'. The key step in this
route involves the treatment of the ate complex XXXIV with a
suitable electrophilic reagent such as tri-n-butyltin chloride,
which induces the preferential migration of the (E)-alkenyl
group (with retention of configuration) from boron to the
adjacent alkynyl carbon atom to produce the dienylborane XXXV.
Presumably initial complexation of the electrophile with the
triple bond occurs, followed by intramolecular transfer of the
(E)-alkenyl group. The electrophile must be capable of giving
an intermediate such as XXXV which can undergo protonolysis under
mild conditions. Protonolysis of both the vinyl carbon-boron and
carbon-tin bonds of the resulting intermediate XXXV with excess
acetic acid affords the corresponding E,Z diene in moderate to
good yield. Boron trifluoride etherate can also be used as the
electrophile in the formation of the organoboron intermediate
(cf. XXXV) from the precursor XXXIV (73) but in our hands
tri-n-butyltin chloride usually gives greater stereoselectivity
and higher yields. Since the intermediates for this synthetic
route are two terminal alkynes, the geometry of the product is
reversed by reversing the order in which the alkynes are used.
Thus, the (E,Z)- and the corresponding (Z,E)-diene isomers are
both available from the same intermediates. This method is
reported to fail when the ate complex contains a (Z)-1-alkenyl
group (73). In Negishi's procedure (cf. Figure 15), use of iodine
as the electrophile results in selective intramolecular transfer
of the (E)-1-alkenyl group to the triple bond but in this case the
resulting organoboron intermediate undergoes elimination of
R_2BI to give the (E)-enyne (71).

The method in Figure 17 often gives a product which contains
traces of the corresponding (E,E)-diene. This isomer can be
selectively removed from the (E,Z)-isomer, as described above,
by formation of its Diels-Alder adduct with excess tetracyanoethylene in tetrahydrofuran followed by chromatography on silica
gel (cf. 17,18). Alternatively, the (E,E)-isomer can be removed
in many cases by the selective formation of its crystalline urea
inclusion complex in methanol (cf. 13).

(E,Z)-1,3-Dienes can also be readily prepared, with reasonable stereoselectivity, by routes involving the Wittig reaction
but these have been recently extensively reviewed (2,6,7) and
therefore will not be discussed here.

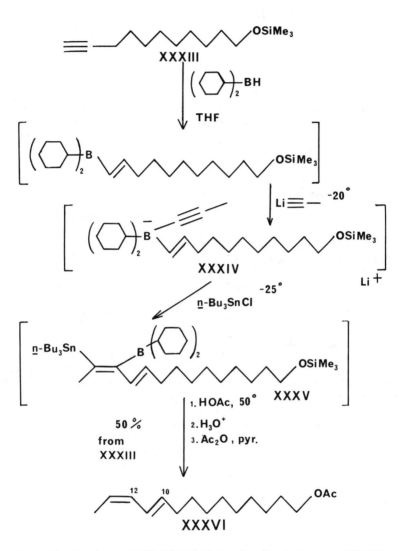

Figure 17. Synthesis of (10E,12Z)-10,12-tetradecadien-1-yl acetate (XXXVI).

(E,E)-1,3-Dienes. The pheromone produced by the female
codling moth, Laspeyresia pomonella, has been shown to be (8E,10E)-
8,10-dodecadien-1-ol (XXXVIII), (74,75). The various synthetic
routes to this compound, which have been reviewed in considerable
detail (2), illustrate some of the methods available for the
synthesis of conjugated E,E dienes (76). Our preferred
route to this compound is outlined in Figure 18 (2,77).
Commercially available sorbyl alcohol (containing 75% of the
E,E isomer) is acetylated and the acetate is coupled with the
Grignard reagent XXXVII in the presence of dilithium tetrachlorocuprate (0.2%) in ether-THF at -20°. This coupling occurs regio-
selectively with no significant allylic rearrangement or change
in the stereochemical purity of the diene system. After preliminary purification, the product is crystallized from pentane
at -10° to give a ≥40% overall yield of pure XXXVIII (>99.5%
E,E isomer). A higher yield is obtained if pure (E,E)-sorbyl
acetate is used as the starting material. Recently several other
workers have published essentially the same route (78,79; see
also 80,81).

An excellent, general procedure for the stereoselective
synthesis of conjugated E,E dienes utilizes 1-halo-1-alkynes
and thexylborane (36). We prepared the compound XXXX (which is
the E,E isomer of the Lobesia botrana pheromone XXIX) in >98%
stereochemical purity by this procedure (Figure 19).

Usually, it is not necessary to use a particularly
stereoselective route to prepare straight chain conjugated (E,E)-
dienols such as XXXVIII and XXXIX. These compounds are, in
general, highly crystalline, which allows their efficient
separation from mixtures by low temperature crystallization from
a hydrocarbon solvent. In addition, any one of the four conjugated diene isomers (or any mixture of these isomers) can be
readily equilibrated by heating without solvent with 1% by weight
of benzenethiol at 100° for 30 minutes (82) to give a mixture
containing ca. 60% of the E,E isomer. Crystallization of this
mixture from pentane at low temperature usually affords the pure,
crystalline E,E isomer and a mixture of isomers in the mother
liquor which can be re-equilibrated. This repetitive equilibration-crystallization procedure allows the efficient conversion
of the other conjugated isomers to the pure E,E isomer.

Thus all four geometrical isomers of 1,4-disubstituted
1,3-dienes can be readily prepared starting from two terminal
acetylene units. The Z,Z isomer is available via Cadiot-
Chodkiewicz coupling (e.g. Figure 12); both the Z,E and E,Z
isomers can be prepared via the procedure in Figure 17; and the
E,E isomer can be isolated from the equilibration-crystallization
procedure described above.

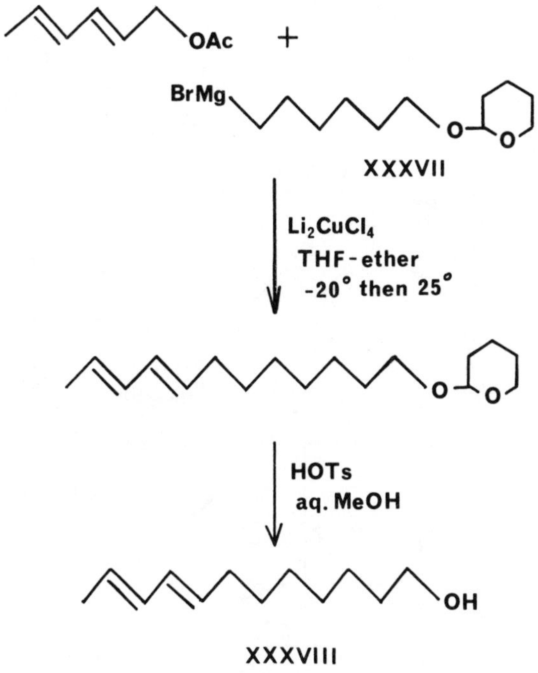

Figure 18. Synthesis of (8E,10E)-8,10-dodecadien-1-ol (XXXVIII).

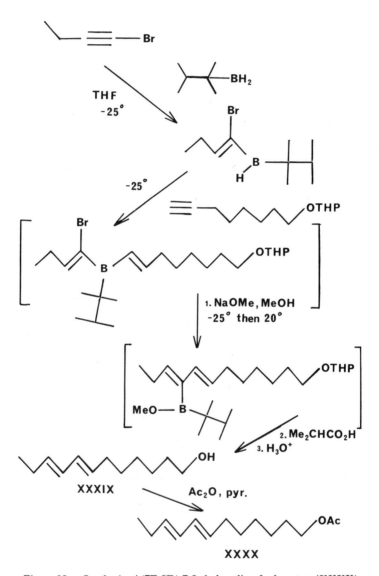

Figure 19. Synthesis of (7E,9E)-7,9-dodecadien-1-yl acetate (XXXX).

Acknowledgments

We gratefully acknowledge the invaluable assistance of our colleagues A. Lui and S. Woo. We wish to especially thank W. Roelofs, M. Gieselmann and their colleagues at the New York State Agricultural Experiment Station in Geneva, N.Y., for extensive collaborative efforts over the past decade.

Literature Cited

1. Katzenellenbogen, J. A. Science 1976, 194, 139.
2. Henrick, C. A. Tetrahedron 1977, 33, 1845.
3. Rossi, R. Synthesis 1977, 817.
4. Rossi, R. Synthesis 1978, 413.
5. Brand, J. M.; Young, Y. C.; Silverstein, R. M. Fortschr. Chem. Org. Naturstoffe 1979, 37, 1.
6. Bestmann, H. J.; Vostrowsky, O. In "Chemie der Pflanzenshutz und Schädlingsbekämpfungsmittel" (R. Wegler ed.)., 1981, 6, 29, Springer Verlag, Berlin.
7. Mori, K. In "The Total Synthesis of Natural Products" (J. ApSimon ed.), 1981, 4, 1, John Wiley and Sons, New York, N.Y.
8. Schlenk, H. J. Am. Oil Chem. Soc.1961, 38, 733.
9. Privett, O. S. In "Progress in the Chemistry of Fats and Other Lipids" 1968, 9 (part 3) Chapt 11, 407, Pergamon Press, Oxford.
10. Schlenk, H. In "Progress in the Chemistry of Fats and Other Lipids" 1954, 2, 243, Pergamon Press, Oxford.
11. Butenandt, A.; Hecker E. Angew. Chem. 1961, 73, 349.
12. Butenandt, A.; Hecker, E.; Hopp, M.; Koch, W. Liebigs Ann. Chem. 1962, 658, 39.
13. Truscheit, E.; Eiter, K. Liebigs Ann. Chem. 1962, 658, 65.
14. Leadbetter, G.; Plimmer, J. R. J. Chem. Ecol. 1979, 5, 101.
15. Mock, W. L.; J. Am. Chem. Soc. 1975, 97, 3666.
16. Nesbitt, B. F.; Beevor, P. S.; Cole, R. A.; Lester R.; Poppi, R. G. Tetrahedron Letters 1973, 4669.
17. Hall, D. R.; Beevor, P. S.; Lester, R.; Poppi, R. G.; Nesbitt, B. F. Chem. Ind. 1975, 216.
18. Goto, G.; Shima, T.; Masuya, H.; Masuoka, Y.; Hiraga, K. Chem. Letters 1975, 103.
19. (a) Gutmann, H.; Lindlar H. In "Chemistry of Acetylenes" (H. G. Viehe, Ed.) 1969, 355, Marcel Dekker, New York, N. Y.; (b) Lindlar, H.; Dubuis, R. Org. Synth. 1966, 46, 89; (c) Marvell, E. N.; Li, T. Synthesis 1973, 457.
20. (a) Rossi, R. Chim. Ind. (Milano)1978, 60, 652; (b) Rossi, R.; Carpita, A.; Gaudenzi, L.; Quirici, M. G., Gazz. Chim. Ital. 1980, 110, 237.
21. (a) Dobson, N. A.; Eglinton, G; Krishnamurti, M.; Raphael, R. A.; Willis, R. G. Tetrahedron 1961, 16, 16 (b) Steenhock, A.; Van Wijngaarden, B. H.; Pabon, H. J. J. Recl. Trav. Chim. Pays-Bas. 1971, 90, 961.

22. (a) Chisholm, M. D.; Steck, W. F.; Underhill, E. W. J. Chem.
     Ecol. 1978, 4, 657; (b) Chisholm, M.D.; Steck, W.; Underhill,
     E.W.; J. Chem. Ecol. 1980, 6 203.
23. Doolittle, R. E.; Proveaux, A. T.; Heath, R. R. J. Chem. Ecol.
     1980, 6, 271; these authors used 5% Pd-on-CaCO₃-quinoline and
     not Lindlar catalyst as stated in their paper (private
     communication).
24. Cram, D. J.; Allinger, N. L., J. Am. Chem. Soc. (1956) 78,
     2518.
25. Brown, C. A.; Ahuja, V. K. J. Chem. Soc. Chem. Commun. 1973,
     553; Brown, C. A.; Ahuja, V. K., J. Org. Chem. 1973, 38, 2226.
26. Roelofs, W. L.; Kochansky, J. P.; Cardé, R. T.; Henrick,
     C. A.; Labovitz, J. N.; Corbin, V. L. Life Sci. 1975, 17, 699.
27. Gunstone, F. D.; Jie, M. L. K. Chem. Phys. Lipids 1970, 4, 1.
28. Morris, S. G.; Herb, S. F.; Magidman, P.; Luddy, F. E.;
     J. Am. Oil Chem. Soc. 1972, 49, 92.
29. Garanti, L.; Marchesini, A.; Pagnoni, U. M.; Trave, R. Gazz.
     Chim. Ital. (1976) 106, 187.
30. Brown, H. C.; Zweifel, G. J. Am. Chem. Soc. 1961, 83, 3834.
31. Fyles, T. M.; Leznoff, C.C.; Weatherston, J. Can. J. Chem.
     1977, 55, 4135.
32. Campbell, K. N.; Eby, L. T. J. Am. Chem. Soc. 1941, 63, 216.
33. Johnson, W. S.; Jensen, N. P.; Hooz, J.; Leopold, E. J.
     J. Am. Chem. Soc. 1968, 90, 5872.
34. Warthen, Jr., J. D.; Jacobson, M., Synthesis 1973, 616.
35. Fyles, T. M.; Leznoff, C. C.; Weatherston, J. J. Chem. Ecol.
     1978, 4, 109.
36. Negishi, E.; Yoshida, T. J. Chem. Soc. Chem. Commun. 1973,
     606.
37. Boeckman, Jr., R. K.; Thomas, E. W. J. Am. Chem. Soc.
     1977, 99, 2805.
38. (a) Slaugh, L, H. Tetrahedron 1966, 22, 1971; Magoon E. F.;
     Slaugh, L. H. Tetrahedron 1967, 23, 4509; (b) Rossi, R.;
     Conti, C. Chim. Ind. (Milano) 1975, 57, 581; Rossi, R.;
     Carpita, A. Synthesis 1977, 561; Rossi, R.; Carpita, A.
     Tetrahedron 1977, 33, 2447.
39. Ashby, E. C.; Lin. J. J.; Goel, A. B. J. Org. Chem. 1978,
     43, 757.
40. Näf, F.; Decorzant, R.; Thommen, W.; Willhalm., B.; Ohloff,
     G. Helv. Chim. Acta 1975, 58, 1016; Winter, M.; Näf, F.;
     Furrer, A.; Pickenhagen, W.; Giersch, W.; Meister, A.;
     Willhalm, B., Thommen, W.; Ohloff, G. Helv. Chim. Acta 1979,
     62, 135.
41. Jacobson, M.; Redfern, R. E.; Jones, W. A.; Aldridge, M. H.
     Science 1970, 170, 542; see also Su, H. C. F.; Mahany, P. G.;
     Brady, U. E. J. Econ. Entomol. 1973, 66, 845.
42. Brown, C. A.; Yamashita, A. J. Am. Chem. Soc. 1975, 97, 891;
     Lindhoudt, J. C.; van Mourik, G. L.; Pabon, H. J. J.
     Tetrahedron Letters 1976, 2565; Brown, C. A.; Yamashita, A.
     J. Chem. Soc. Chem. Commun. 1976, 959; Hommes, H.; Brandsma,
     L. Recl. Trav. Chim. Pays-Bas. 1977, 96 160.

43. Brandsma, L. "Preparative Acetylenic Chemistry", 1971, p 30, Elsevier, New York, N.Y.; Danehy, J. P.: Killian, D. B.; Nieuwland, J. A. J. Am. Chem. Soc. 1936, 58, 611; Leznoff, C.C.; Sondheimer, F. J. Am. Chem. Soc. 1967, 89, 4247; Heslinga, L.; Pabon, H. J. J.; Van Dorp, D. A. Recl. Trav. Chim. Pay-Bas. 1973, 92, 287.
44. Sevin, A.; Chodkiewicz, W.; Cadiot, P. Bull. Soc. Chim. Fr. 1974, 913
45. Castro, C. E.; Haulin, R.; Honwad, V. K.; Malte, A.; Mojé, S. J. Am. Chem. Soc. 1969, 91, 6464.
46. Goto, G.; Masuoka, Y.; Hiraga, K. Chem. Letters 1974, 1275.
47. Roelofs, W. L.; Gieselmann, M. J.; Cardé, A. M.; Tashiro, H.; Moreno, D. S.; Henrick, C. A.; Anderson, R. J. Nature 1977, 267, 698.
48. Roelofs, W.; Gieselmann, M.; Cardé, A.; Tashiro, H.; Moreno, D. S.; Henrick, C. A.; Anderson, R. J. J. Chem. Ecol. 1978, 4, 211.
49. Anderson, R. J.; Adams, K. G.; Chinn, H. R.; Henrick, C. A. J. Org. Chem. 1980, 45, 2229.
50. Gieselmann, M. J.; Henrick, C. A.; Anderson, R. J.; Moreno, D. S.; Roelofs, W. L. J. Insect Physiol. 1980, 26, 179.
51. Still, W. C.; Mitra, A., J. Am. Chem. Soc. 1978, 100, 1927.
52. Henrick, C. A.; Anderson, R. J.; Carney, R. L. In "Regulation of Insect Development and Behavior" (F. Sehnal, A. Żabża; J. J. Menn; B. Cymborowski, scientific eds.), 1981, Part II, 887, Wrocław Technical University Press, Wrocław, Poland.
53. For an alternative synthesis of XV see Snider, B. B.; Rodini, D. Tetrahedron Letters 1978, 1399.
54. Henrick, C. A.; Anderson, R. J.; Staal, G. B.; Ludvik, G. F., J. Agric. Food Chem. 1978, 26, 542.
55. Gieselmann, M. J.; Rice, R. E.; Jones, R. A.; Roelofs, W. L. J. Chem. Ecol. 1979, 5, 891.
56. Anderson, R. J.; Chinn, H. R.; Gill, K.; Henrick, C. A. J. Chem. Ecol. 1979, 5, 919.
57. Anderson, R. J.; Gieselmann, M. J.; Chinn, H. R.; Adams, K. G.; Henrick, C. A.; Rice, R. E.; Roelofs, W. L. J. Chem. Ecol. 1981, 7, 695.
58. Heath, R. R.; McLaughlin, J. R.; Tumlinson, J. H., Ashley, T. R.; Doolittle, R. E. J. Chem. Ecol. 1979, 5, 941.
59. Heath, R.R.; Doolittle, R. E.; Sonnet, P. E.; Tumlinson, J. H. J. Org. Chem. 1980, 45, 2910.
60. Zweifel, G.; Polston, N. L. J. Am. Chem. Soc. 1970, 92, 4068.
61. Cadiot, P., Chodkiewicz, W. In "Chemistry of Acetylenes" (H. G. Viehe, ed.) 1969, Chapt 9, 597, M. Dekker, New York, N.Y.
62. Pelter, A.; Hughes, R.; Smith, K.; Tabata, M. Tetrahedron Letters 1976, 4385.
63. Sinclair, J. A.; Brown, H. C. J. Org. Chem. 1976, 41, 1078.
64. Carney, R. L.; Henrick, C. A. 1980, U.S. Patent 4,198,533.

65. Coffelt, J. A.; Vick, K. W.; Sonnet, P. E.; Doolittle, R. E.
    J. Chem. Ecol. 1979, 5, 955.
66. Sonnet, P. E.; Heath, R. R. J. Chem. Ecol. 1980, 6, 221.
67. For some additional general methods for the preparation of
    1,3-enynes and/or (E,Z)-1,3-dienes see also Baba, S.;
    Negishi, E., J. Am. Chem. Soc.1976, 98, 6729; Dang, H. P.;
    Linstrumelle, G. Tetrahedron Letters 1978, 191; Miyaura, N.;
    Yamada, K.; Suzuki, A. Tetrahedron Letters 1979, 3437;
    Commercon, A.; Normant, J. F.; Villieras, J. Tetrahedron
    1980, 36 1215; Cassani, G.; Massardo, P.; Piccardi, P.
    Tetrahedron Letters 1979, 633; Jabri, N.; Alexakis, A.;
    Normant, J. F. Tetrahedron Letters 1981, 959; Brown, H. C.;
    Molander, G. A. J. Org. Chem. 1981 46, 645; Miyaura, N.;
    Suginome, H.; Suzuki, A. Tetrahedron Letters 1981, 127.
68. Labovitz, J. N.; Graves, V. L.; Henrick, C. A. 1974, un-
    published results, Zoecon Corporation.
69. Labovitz, J. N.; Henrick, C. A.; Corbin, V. L. Tetrahedron
    Letters 1975, 4209.
70. Many of the routes to XXIX have been recently reviewed in
    ref 2; for more recent syntheses see (a) Descoins, C.;
    Samain, D.; Lalanne-Cassou, B.; Gallois, M. Bull. Soc. Chim.
    Fr. 1977, 941; (b) Bestmann, H. J.; Süss, J.; Vostrowsky,
    O. Tetrahedron letters 1979, 2467; (c) Cassani, G.; Massardo,
    P.; Piccardi, P. Tetrahedron Letters 1980, 3497; Dressaire,
    G.; Langlois, Y. Tetrahedron Letters 1980,67; Ratovelomanana,
    V.; Linstrumelle, G. Tetrahedron Letters 1981, 315.
71. Negishi, E.; Lew, G.; Yoshida, T. J. Chem. Soc. Chem. Commun.
    1973, 874.
72. Negishi, E.; Abramovitch, A. Tetrahedron Letters 1977, 411.
73. Zweifel, G.; Backlund, S. J. J. Organomet. Chem. 1978, 156,
    159.
74. Roelofs, W. L.; Comeau, A.; Hill, A.; Milicevic, G.; Science
    1971, 174, 297.
75. Beroza, M.; Bierl, B. A.; Moffitt, H. R. Science 1974, 183,
    89; McDonough, L. M.; Moffitt, H. R. Science 1974, 183, 978.
76. For an additional general method for the preparation of
    (E,E)-1,3-dienes see Okukado, N.; VanHorn, D. E.; Klima,
    W. L.; Negishi, E. Tetrahedron Letters 1978, 1027.
77. Henrick, C. A.; Anderson, R. J.; Rosenblum, L. D. 1974,
    unpublished results, Zoecon Corporation.
78. Bestmann, H. J.; Süss, J.; Vostrowsky, O. Tetrahedron
    Letters 1978, 3329.
79. Samain, D.; Descoins, C.; Commercon, A. Synthesis 1978, 338.
80. Decodts, G.; Dressaire, G.; Langlois, Y. Synthesis 1979, 510.
81. Samain, D.; Descoins, C.; Langlois, Y. Nouv. J. Chim. 1978,
    2, 249.
82. Henrick, C. A.; Willy, W. E.; Baum, J. W.; Baer, T. A.;
    Garcia, B. A.; Mastre, T. A.; Chang, S. M. J. Org. Chem.
    1975, 40, 1.

RECEIVED February 24, 1982.

Chiral Insect Sex Pheromones:
Some Aspects of Synthesis and Analysis

P. E. SONNET and R. R. HEATH

U.S. Dept. of Agriculture, Agricultural Research Service, Insect Attractants,
Behavior, and Basic Biology Research Laboratory, Gainesville, FL 32604

Asymmetric syntheses directed toward construction
of enantiomers of the western and southern corn rootworm pheromones are described. A brief review of the
subject of asymmetric synthesis as it is related to the
synthesis of insect sex pheromones is presented. The
laboratory's previous research with chiral pheromones
is summarized (Japanese beetle, white peach scale, and
lesser tea tortrix) before detailing synthetic work on
the pheromones of the aforementioned rootworm species.
Throughout the course of the synthetic effort, cholesteric stationary phases for GLC have found use. Their
superior ability to separate crucial diastereomeric
intermediates for synthesis is detailed.

Stereochemistry and asymmetric synthesis are topics with
which chemists traditionally have been concerned (1). In recent
years there has been a virtual explosion of literature in the
area of asymmetric organic synthesis that has fortuitously paralleled the increased awareness of insect pheromone stereochemistry. Many useful reviews of asymmetric synthesis exist (2, 3,
4, 5, 6) and this paper will only briefly direct the reader's
attention to examples of reported syntheses by type that may be
of potential general use for pheromone synthesis. It should be
clear even to the casual reader that this field is in need of
almost annual review and current literature would have to be
consulted in the face of an original problem in synthesis.

Asymmetric Synthesis

Since the relationship of biological activity to pheromone
configuration can only be assessed correctly if "pure" compounds
are made available, the research effort must indeed seek to prepare stereospecifically the several stereoisomers implied by

a given structure (7, 8, 9). Most syntheses of chiral pheromones to date have employed a chiral starting material. In such
instances the absolute configuration of the product is ensured;
and providing the enantiomeric excess of the starting material
is known and a route that does not compromise the stereocenters
has been chosen, the configurational purity of the product is
also precisely known. Often, however, a useful chiral starting
material may not be available necessitating a circuitous route.
Additionally, the configurational purity of a commercial material may not have been determined absolutely, i.e., only a maximum
optical rotation is recorded. The investigator must then devise
a method, perhaps at some stage of synthesis, to determine
enantiomeric purity.

Faced with these dilemmas one might consider alternative
methods of generating asymmetry. These may be grouped according
to whether diastereomeric compounds, or complexes, are formed.
Analogous types of resolution are, in principle, possible in
each group. Thus diastereomeric compounds may be fractionally
crystallized or separated by chromatographic means. On the
other hand, transient diastereomeric complexes formed by chiral
stationary phases can also afford chromatographic enantiomer
separation (10, 11, 12). Occasionally enantiomer resolution is
achieved spontaneously by crystallization; an example has been
given by Still (13) in his synthesis of Periplanone-B, sex
pheromone of the American cockroach. In a second broadly
applicable method, kinetic resolution via diastereomers may be
achieved whenever one of a pair preferentially undergoes chemical reaction. Similarly, enantiomers may themselves be kinetically resolved by preferential reaction with a chiral reagent
-- Sharpless has pointed out the unique capacity for such
methodology to provide absolutely pure enantiomers with elegant
exemplification (14). The interconversion of diastereomers via
a stereochemically labile center is sometimes an aid to resolution, and deracemization using a chiral reagent has also been
described (15). Finally, it is possible to induce asymmetry in
a prochiral molecule by reaction with a chiral reagent, or perhaps in a chiral solvent, or quite commonly intramolecularly
with intimate involvement of a preexisting chiral center. The
enantiomeric excess of the created center is determined by the
difference in free energies of the two diastereomeric transition
states leading to the products. For a reaction conducted at
25°C, an enantiomer excess of 99%, implies an energy difference
of 3.1 kcal/mol. Despite the stringent nature of such a
requirement, the difficulty in its achievement, and the conse-
quent problem of further purification, this last method enjoys
great popularity. As some of these reactions adumbrate future
pheromone syntheses, and we have employed some of this chemistry
in our own subsequently described efforts, they are enumerated
below.

Reduction. Successful asymmetric hydrogenations of olefin
double bonds mediated by chiral phosphines have been reported
(16) and the factors crucial for effective asymmetric induction
in related systems have been discussed (17, 18). These reductions require functionality proximate to the double bond for any
degree of success.

Noncatalytic reduction, particularly for the carbonyl group,
has been more successful. Since Mosher (19) first described the
properties of lithium aluminum hydride (LAH) modified by the
chiral aminoalcohol, Darvon alcohol (Figure 1), numerous additional examples and modifications have been published. Of
particular value is Brinkmeyer's observation that high enantiomeric excesses may be obtained using LAH-Darvon when the substrate is an alkynone (20). Reduction of this triple bond provides optically active saturated secondary alcohols. LAH modified by enantiomerically pure 1,1'-bi-2-naphthol has been
employed in syntheses of the pheromones of the Japanese beetle
and dried bean beetle (21). Another very useful reagent is the
borane formed from 9-BBN and α-pinene (Figure 1) (22). Reduction of alkynones is very stereoselective with this reagent, the
reagent from (+)-α-pinene generating the R-alkynol. The
pinanyl borane has been employed in a synthesis of the sex
pheromone of the Japanese beetle (23).

Electrophilic addition. The examples available succeed by
virtue of functionality placed near the reacting center. In the
instances illustrated (Figure 2) the hydroxyl group of an
allylic alcohol participates intimately with the electrophilic
species to provide an ability to discriminate the faces of a
double bond (24, 25). The latter reaction was employed in a
sequence eventuating in the sex pheromone of the gypsy moth,
disparlure.

Electrophilic substitution. A number of chiral nucleophilic
species have been described that result in optically active α-
alkyl aldehydes, ketones, acids, and acid derivatives upon
alkylation and (usually) subsequent hydrolytic cleavage. Enders
provides a number of examples (Figure 3) one of which results in
the ant alarm pheromone, 4-methyl 3-heptanone (26, 27). Studies
by A. I. Meyers of the chemistry of anions of chiral oxazolines
(Figure 4) were the first of the genre, however (28). Related
reactions of chiral anions of metalloenamines and hydrazones
(29, 30, 31) have in common with the alkylation of oxazolines
metallated azaenolate intermediates that predispose one face of
an azaenolate double bond to reaction with the electrophile.
More recently α-C-metallated amides of the chiral amine,
(1)-ephedrine, have been described (32) (Figure 5) as a preparation for α-alkylalkanoic acids. In this case the hydroxyl
substituent has been suggested to produce a chelated nucleo-
phile. Analogous amides of prolinol can be similarly

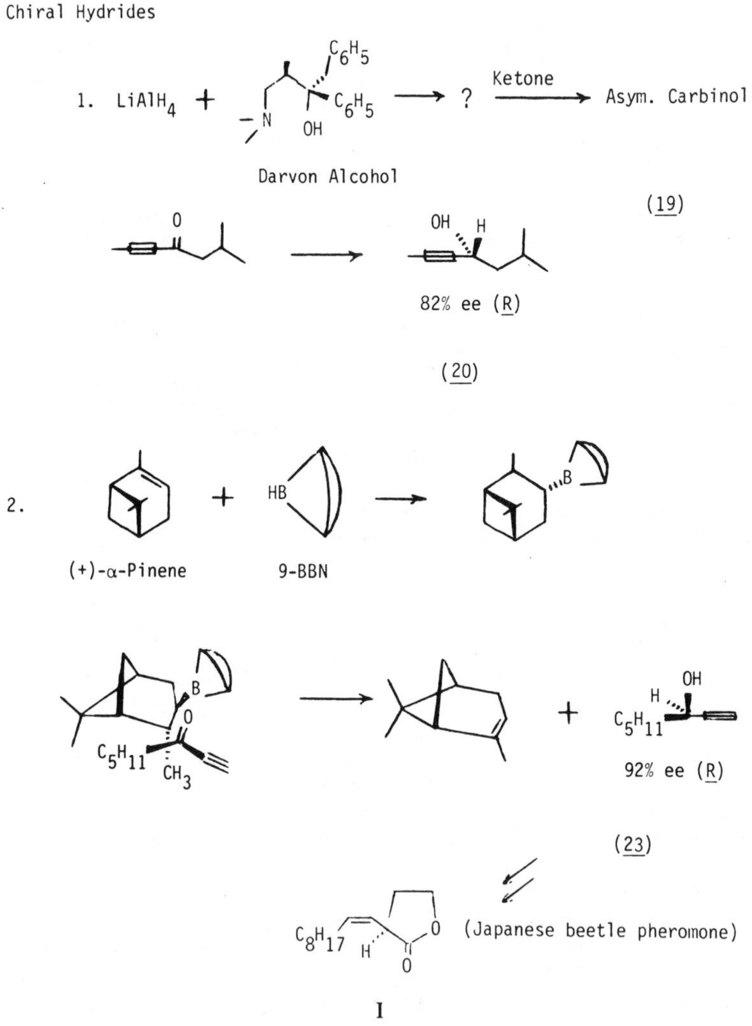

Chiral Hydrides

1. LiAlH$_4$ + [structure of Darvon Alcohol with C$_6$H$_5$ groups, N, OH] → ? $\xrightarrow{\text{Ketone}}$ Asym. Carbinol

Darvon Alcohol

(19)

[structure of ketone] → [structure of carbinol with OH, H]

82% ee (R)

(20)

2. [structure of (+)-α-Pinene] + [structure of 9-BBN, HB] → [structure of product with B]

(+)-α-Pinene   9-BBN

[structure with B, O, C$_5$H$_{11}$, CH$_3$] → [structure of α-pinene] + [structure with H, OH, C$_5$H$_{11}$]

92% ee (R)

(23)

[structure, C$_8$H$_{17}$, H, O, O] (Japanese beetle pheromone)

I

Figure 1.   Noncatalytic asymmetric reduction (19, 20, 23).

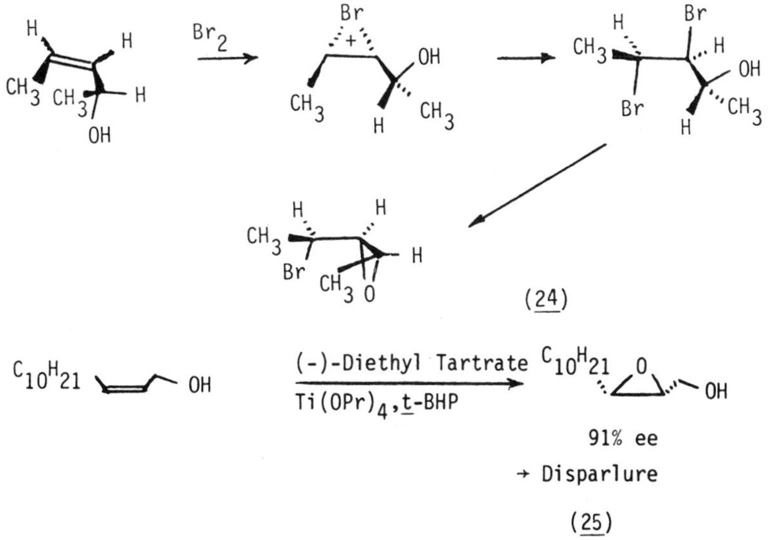

Figure 2.   *Asymmetric electrophilic addition (24, 25).*

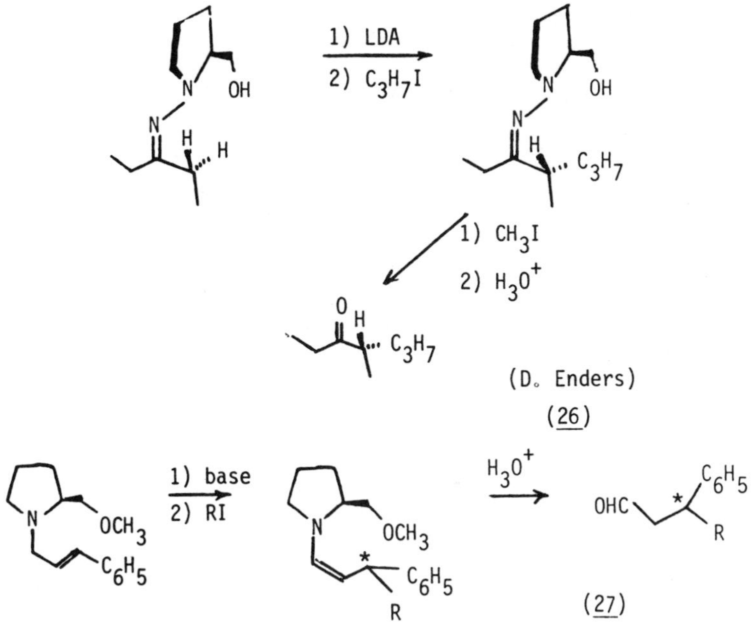

Figure 3.   *Asymmetric electrophilic substitution (26, 27).*

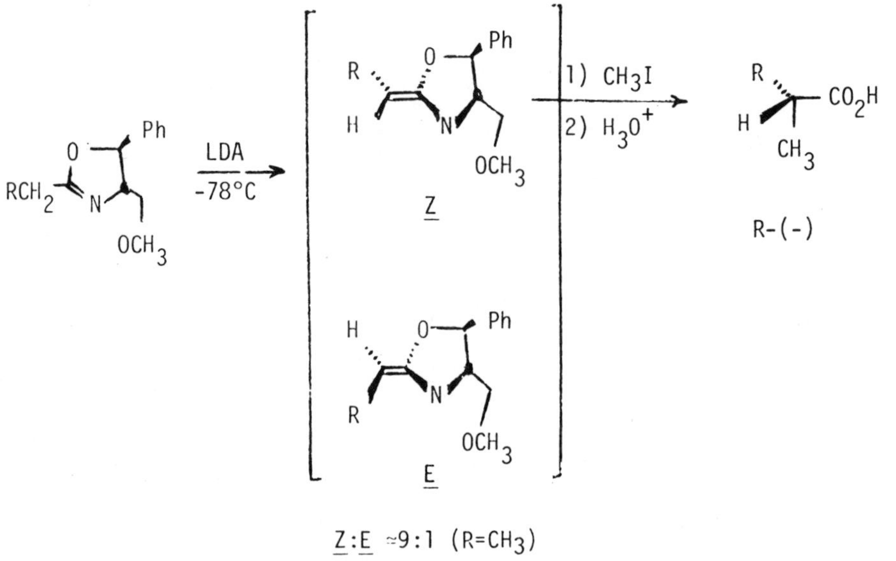

$$\underline{Z}:\underline{E} \approx 9:1 \quad (R=CH_3)$$

$$(\underline{28})$$

Figure 4.   Alkylation of anions derived from chiral oxazolines (28).

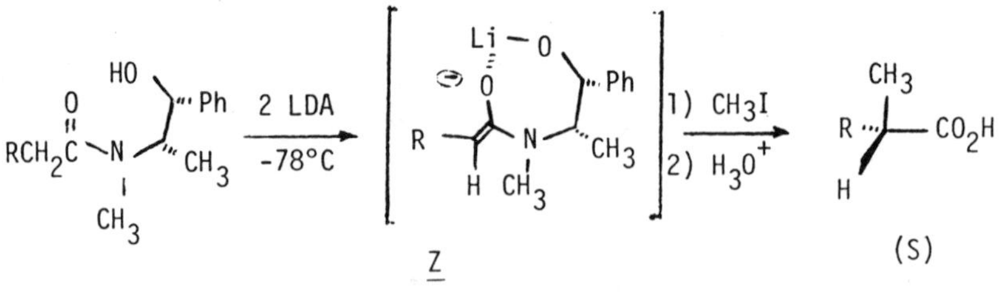

$$(\underline{32})$$

Figure 5.   Alkylation of anions derived from amides of (1)-ephedrine (32).

alkylated (33, 34) and an examination of the aminoenolate intermediate by NMR techniques revealed a single enolate species presumably Z (Figure 6) (33). Moreover, alkylation of prolinol
(and ephedrine) amide anions in which the hydroxyl function is
protected results in induction of asymmetry in the opposite
sense. Some specific data is provided below in the discussion
of our synthetic work. Alkylations such as these set the stage
for synthesizing branched sites in pheromones. In addition,
reactions with 2-carbon and 3-carbon electrophiles have been
employed to prepare alkylated lactones (35). Other recent
examples of the alkylations of chiral enolates, exclusive of
aldol condensations, are provided in the following references:
(36-40).

Syntheses of Insect Sex Pheromones

Japanese beetle. The synthesis of this pheromone, (R,Z)-5-
(1-decenyl) dihydro-2(3H)-furanone (Figure 1), I, has already
been described and need not be repeated here (41). The R-acid
was a key intermediate in this synthesis (Figure 7), and was
obtained by nitrosation of R-glutamic acid. The enantiomeric
glutamic acids can be obtained commercially and in high purity
although only the S-isomer is relatively inexpensive. No less
than seven pheromones have been synthesized via the R and S lactone acid demonstrating the great versatility of this intermediate and the utility of glutamic acid as a chiral starting
material (42).

The pure synthetic R,Z isomer was a powerful attractant for
male Japanese beetles in field bioassays, whereas the S,Z isomer
was a strong inhibitor. As little as 1% of the S,Z-isomer
inhibited male response.

White peach scale. Several scale sex pheromones have now
been elucidated; each of them possesses an asymmetric center and
usually a trisubstituted alkene link within an isoprenoid framework (43). The structure of the white peach scale pheromone,
R,Z-II (Figure 8), lent itself to synthesis with another chiral
starting material, namely limonene (44). Selective ozonlysis
followed by workup with dimethyl sulfide-methanol provided a
ketoacetal, III. Wittig methylenation followed by hydrolytic
cleavage of the acetal gave a dienaldehyde, IV. Conversion of
the aldehyde via the acid to an amide (45) with enantiomerically
pure α-methylbenzylamine permitted chromatographic assessment
of the purity of the diene aldehyde (and the limonene). The
required R-isomer of the diene aldehyde was >98% ee.

Although the trisubstituted olefinic linkage of the pheromone could be fashioned readily in a less selective manner followed by HPLC purification, we sought a stereoselective route in
addition. Still and Mitra completed a synthesis of the

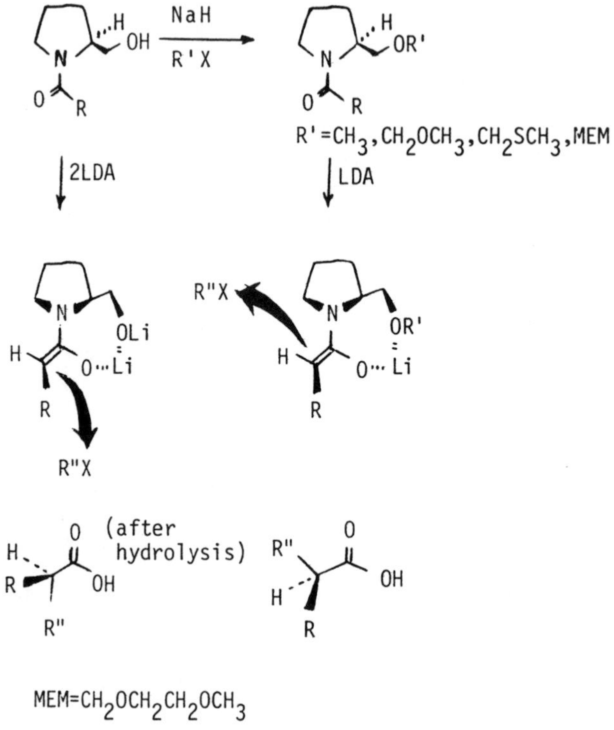

Figure 6.   *Predominant configurations obtained by reaction of the chiral enolates of an amide of (S)-(−)-prolinol and its O-alkyl derivatives with ethyl iodide are shown. Final products are acid obtained by hydrolyzing the alkylated amide.*

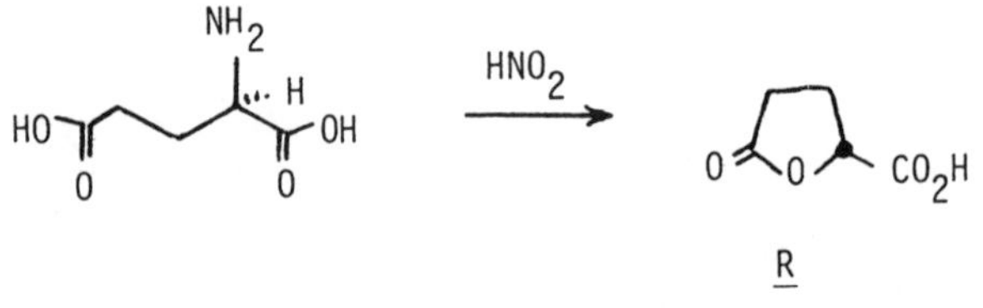

Figure 7.   *Pheromones from R or S lactone acid for beetles: dermestids (γ-capro-lactone),* Pityogenes chalcographus *(L.), Japanese beetle (I, Fig. 1), ambrosia beetle, and lesser grain borer beetle; gypsy moths; and black-tailed deer.*

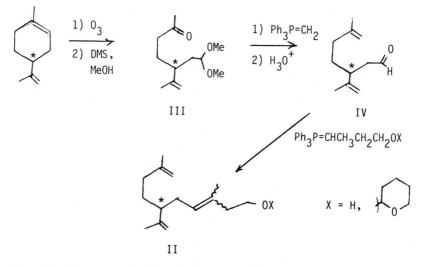

Figure 8.   Limonene as a chiral starting material for white peach scale phero-mone.

California red scale pheromone using a (2,3)-sigmatropic rearrangement of a tin-derived methyl allyl ether anion that provided the Z geometry also required for the white peach scale
pheromone (46). We examined the SCOOPY reaction, a family of
modified Wittig reactions employed initially by Christmann and
Schlosser for E-olefins (47). The betaines derived from
reactions of R and S dienealdehydes with a suitable 3-carbon
chain phosphorane (Figure 9) were deprotonated and allowed to
react with methyl iodide. Hydrolysis of the products gave the
E-alkenes (35% yield of 97% E). The corresponding Z-alkenes
were obtained by an analogous sequence involving ethylidene triphenylphosphorane and ethylene oxide. This sequence provided
the Z-homoallylic structures in a disappointing 15% yield (98%
Z).

Only the R,Z-stereoisomer elicits sexual response from male
white peach scale, the other isomers have no biological activity
nor do they affect the activity of the R,Z-isomer.

Lesser tea tortrix. A minor component of the sex pheromone
blend of this moth is 10-methyl-1-dodecanol acetate, V, (Figure
10) (48). Mori has synthesized the enantiomers; both compounds
were constructed from (R)-(+)-citronellol (49). The racemic
target compound was synthesized by alkylating 10-undecenoic acid
(as its dianion) with ethyl iodide (50). Reduction of the
carboxyl to a methyl group was accomplished by standard procedures. Hydroboration of the olefinic link with disiamylborane
and oxidative workup yielded the primary alcohol which was then
acetylated to give the racemic pheromone structure. The overall
yield from undecenoic acid was 40-45% without any extensive
effort to optimize.

Modifications to this route were devised to permit some
latitude in obtaining configurationally biased products.
Undecenoic acid was converted to an amide of (S)-(-)-prolinol
(Figure 11), VI. The MEM ether of this amide was prepared with
sodium hydride-THF followed by MEM-chloride. Other O-deriva-
tives were also prepared, and a number of trial alkylations were
conducted (LDA as base, ethyl iodide as alkylating agent). It
was possible to obtain the amide with a 76% bias in the acid
residue for S-configuration when the hydroxyl group of the
chiral auxiliary was present as "OLi," and a 74% R-ee was
obtainable by alkylating the MEM derivative. The MEM ether link
was cleaved with ZnBr$_2$, and hydrolytic cleavage of the prolinol amides was then conducted in a two-phase system composed
of hexane and 2N HCl under reflux (4-6 h). Further purification
of the alkylated prolinol amides by HPLC was also affected, producing (after hydrolysis) α-ethylated acids that were 94-96%
ee. These acids were then transformed as previously described
(Figure 10).

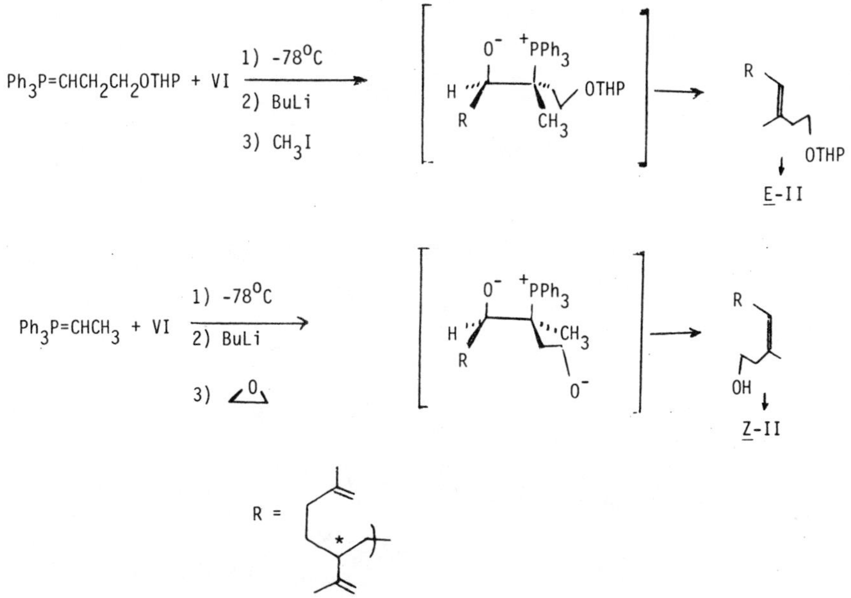

Figure 9. *Use of betaine anions to construct a required tri-substituted homoallylic alcohol.*

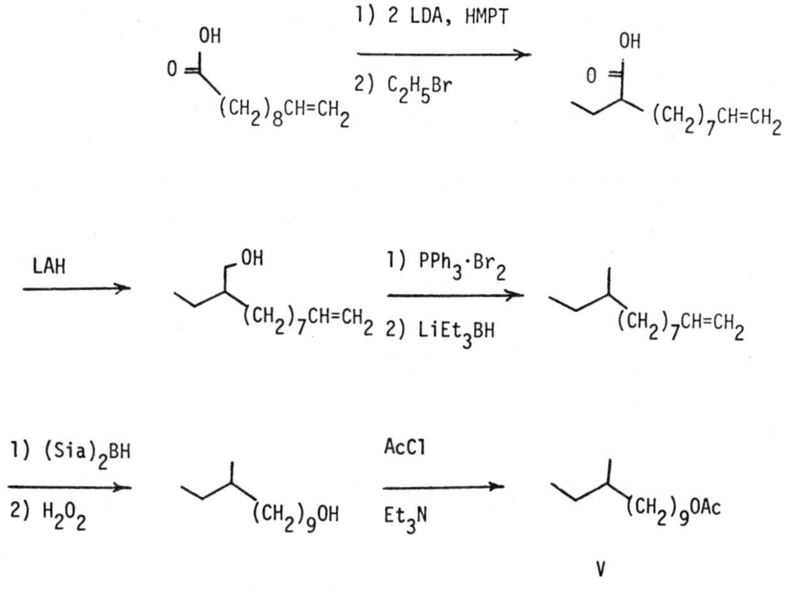

Figure 10. *Synthesis of (+)-10-methyl-1-dodecanol acetate.*

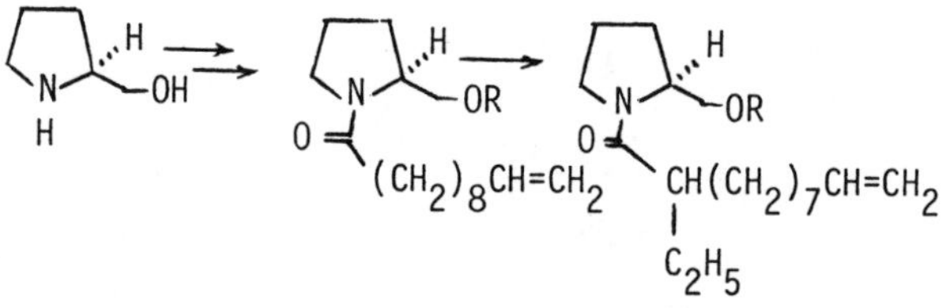

VI

R	Induced·Configuration
H	S,76% ee
MEM	R,74% ee

Figure 11.   Alternative preparations of diastereomeric amides.

Western corn rootworm. The sex pheromone of the western
corn rootworm has been identified as 8-methyl-2-decanol propanoate (51), VII (Figure 12). This appears to be the first identification within the coleopteran family Chrysomelidae. Initially
we synthesized the 2(R), 8(R,S) and 2(S), 8(R,S) structures with
the supposition that chirality at the oxygenated function might
be more crucial than at the hydrocarbon end of the molecule.
Moreover, it permitted us to address questions of asymmetric
induction in generating the configuration at C-2. The features
of interest in this route are hydrogenation of homoallylic
bromide, VIII, apparently uncomplicated by hydrogenolysis, and
preparative HPLC collections of MTPA (52) esters of the alkynols
IX. Although such derivatives are well established for these
purposes, the saturated diastereomers could not be easily
purified. The HPLC-collected derivatives could be hydrogenated
as MTPA esters without complication.

Reduction of akynones to alkynols as discussed earlier is
potentially a very important reaction for insect pheromone synthesis. Accordingly 8-methyl-3-decyn-2-one, X (Figure 13), a
synthetic precursor of the western corn rootworm pheromone, was
subjected to reduction with LAH modified by several chiral
aminoalcohols. Reductions with (1)-ephedrine (secondary amines
appeared as yet not to have been exploited as ligands) were comparable to those reported for Darvon (20). Thus, LAH/1ephedrine
prepared as first described for LAH/Darvon (19) and allowed to
react with the alkynone at 25°C produced 68% ee (R)-alkynol (88%
reduction); reduction at -100°C gave 80% ee (R) ($\overline{7}0$% reduction). Using these latter conditions 2-octanone only yielded
10% ee. Reductions based on LiBH$_4$ (apparently also not previously reported) were much poorer than LAH although one might
have hoped that the shorter bond lengths involving boron would
have intensified diastereomer energy differences in the reduction transition complex.

Optimum asymmetric reductions of alkynones previously had
been achieved employing the chiral legand in a 2:1 ratio with
LAH. Dimeric ligands were synthesized from (1)-ephedrine and
(S)-(-)-prolinol and employed 1:1 with LAH. The results (using
a 3-carbon bridge) have not been encouraging (Figure 13).

We turned our efforts to a synthesis in which a chiral
5-carbon unit would be coupled to a 6-carbon structure bearing
functionality permitting resolution. The 5-carbon fragment
would contain the hydrocarbon asymmetric center (C-8 of VII);
the other unit would provide C-2. Commercially available
(S)-2-methyl 1-butanol was determined to be >99% pure. However, the R-alcohol (acid, aldehyde, etc.) would have to be
synthesized. Asymmetric alkylations of chiral α-metallated
amides were performed, but the enantiomeric excesses were not
sufficiently high. In particular we noted that alkylations
involving a short chain bifunctional compound (e.g., 3-methoxy-
propyl iodide) provided slightly lower ee's than did the parent
alkyl iodide.

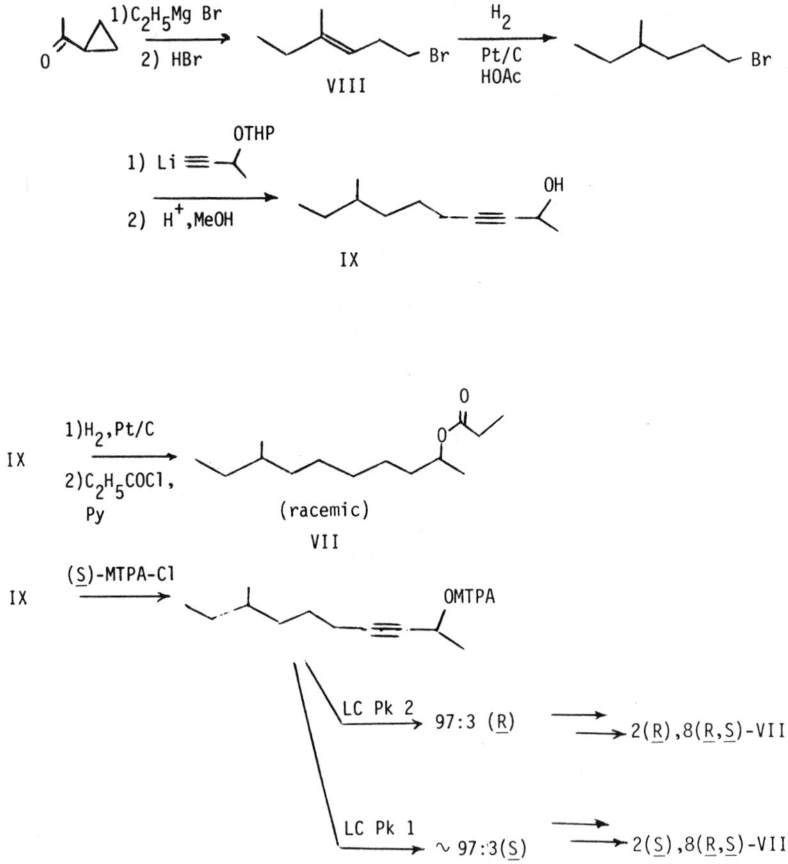

Figure 12.   Syntheses of 2(R), 8(R,S) and 2(S), 8(R,S) structures.

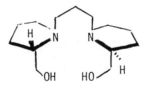

(**H**)*

LAH/1-ephedrine (<u>19</u>),25°C   (<u>R</u>)-68% ee

−100°C   (<u>R</u>)-80% ee

C-2 configuration

LAH, THF-Ether, Mosher's Procedure A   ∿ 2% ee (<u>S</u>)

LAH, THF-Ether, Mosher's Procedure B (6h)   30% ee (<u>S</u>)

LAH, THF-Ether, Mosher's Procedure A   14% ee (<u>R</u>)

LAH, THF-Ether, Mosher's Procedure B (6h)   22% ee (<u>R</u>)

Figure 13.   Reductions of alkynones with some chiral hydrides.

R. Carney, of Zoecon Corporation, Palo Alto, Calif., who is
collaborating in this research, developed a synthesis of 96% ee
(R)-2-methylbutyric acid from D-isoleucine (Figure 14). Another
approach that involved HPLC purification of diastereomeric
amides with subsequent cleavage of the amide link gave the
R-acid in 94% ee.

The chiral 5-carbon units were now in hand as was a route
that would permit HPLC resolution of C-2 stereoisomers (Figure
12); the synthesis was concluded as outlined in Figure 15.

Southern corn rootworm. The structure of the southern corn
rootworm has been defined as 10-methyl 2-tridecanone, XI (Figure
16) (53). Alkylation of undecanoic acid with n-propyl bromide
was followed by conversion to the diastereomeric amides with
either (S)- or (R)-α-methylbenzylamine that had been purified
previously by recrystallization of D and L tartaric acid salts,
respectively. Recrystallizations of these amides from ethanol
(4 was sufficient) gave 32% yields of pure (>99.5%) diastereomers (Figure 16). Hydroxyethylation labilized the amides toward
hydrolysis. It was convenient to intercept the aminoesters and
reduce them with LAH. The resulting carbinols were than carried
forward in standard manner to provide the ketones.

Analyses of Synthetic Intermediates

Throughout this research we have made constant use of
columns coated with cholesteric phases (54-55). Thus, for
example, asymmetric induction in amide alkylations as well as
asymmetric reductions of carbonyl were conveniently judged
either on the product directly (amides) or on any of several
derivatives of the product (alcohols) with cholesteryl cinnamate
as a GLC stationary phase.

Many studies of separations, primarily by HPLC, have been
reported with a view to establishing the mechanisms by which
diastereomers separate (56-58). Although substantive criticism
has been offered by way of demonstrated exceptions (45, 59), the
picture as qualified by Pirkle (57) holds well for simple
(otherwise functionally unsubstituted) amides and carbamates.
Briefly, solution conformations of amides and carbamates are as
shown in Figure 17; the central functional group acts to create
a plane, and the asymmetric centers extend alkyl (aryl) residues
to either side of that plane. An explanation based on a combination of steric bulk and hydrophobicity has been advanced to
explain HPLC elution orders for carbamates. Elution order for
GLC is inverted from that of HPLC in all instances studied, the
least soluble (faster eluting) diastereomer by GLC is always the
cisoid molecule. The notable exceptions are those compounds
bearing a CF_3 group at R_1.

Our primary interest, however, was to assess the relative
separation efficiencies of several liquid phases versus

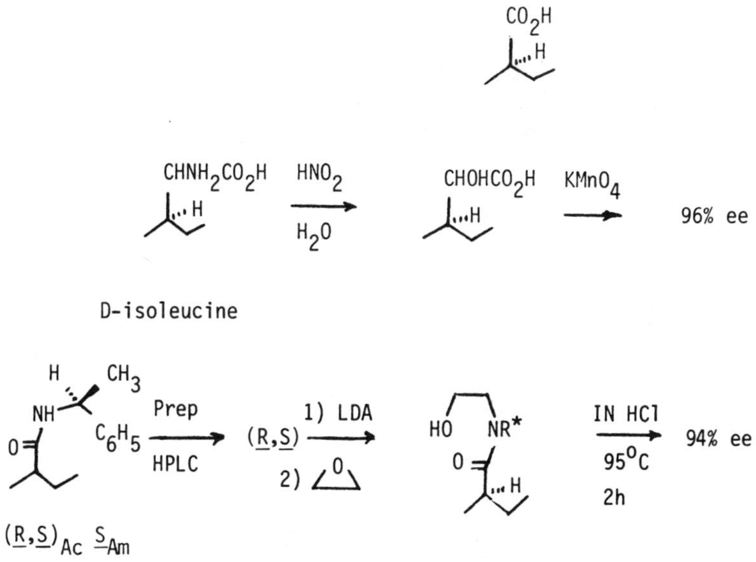

Figure 14. *Alternative syntheses of the 5-carbon chiral unit, (R)-2-methylbutyric acid.*

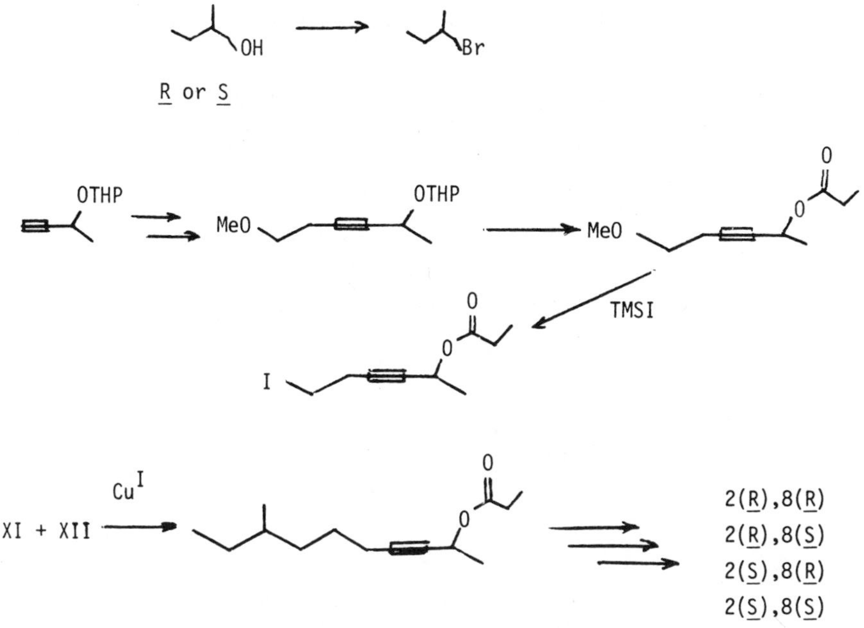

Figure 15. *Synthesis of the enantiomers of the western corn rootworm sex pheromone.*

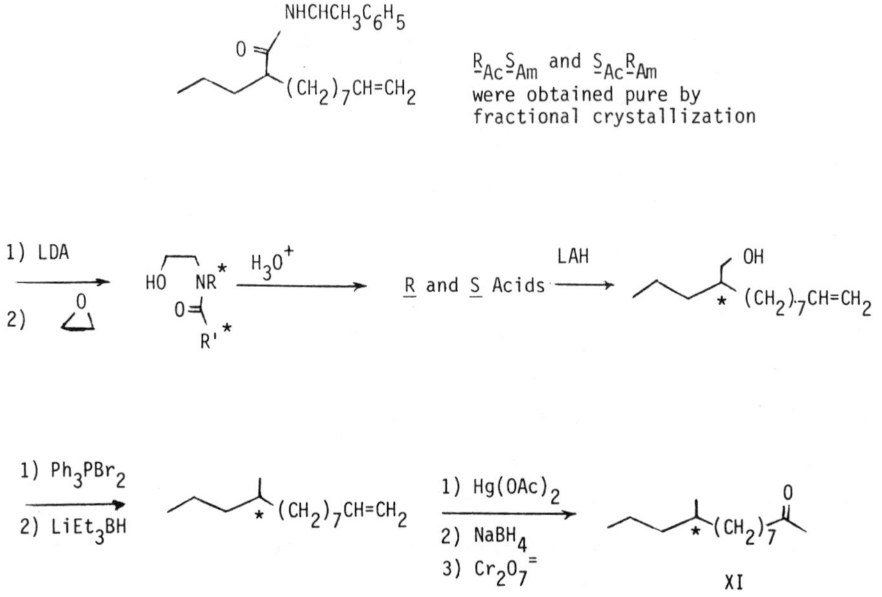

$$\underline{R}_{Ac}\underline{S}_{Am} \text{ and } \underline{S}_{Ac}\underline{R}_{Am}$$
were obtained pure by
fractional crystallization

Figure 16.   Synthesis of the enantiomers of the southern corn rootworm sex pheromone.

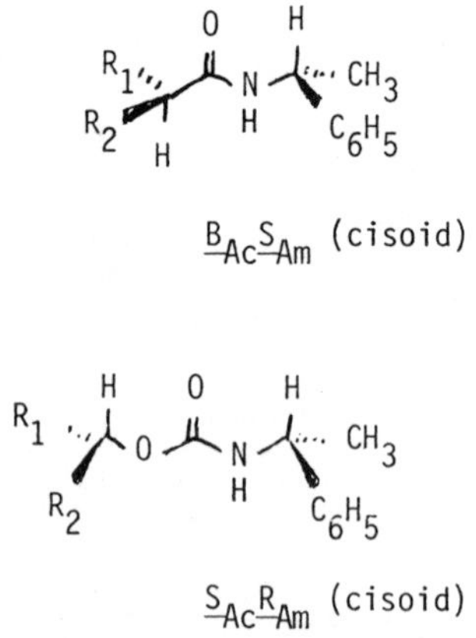

$\underline{B}_{Ac}\underline{S}_{Am}$ (cisoid)

$\underline{S}_{Ac}\underline{R}_{Am}$ (cisoid)

Figure 17.   Solution conformers of diastereomeric amides and carbamates. R_1 and
R_2 are saturated alkyl groups; R_2 is longer than R_1.

cholesteryl para-chlorocinnamate (CpCC), one of several
cholesteric phases currently under evaluation in our laboratory. As R_1 or R_2 were aryl in previously reported investigations, and our own investigations required separations of
amides bearing alkly substituents, we prepared and analyzed a
number of diastereomeric pairs in which R_1 and R_2 were
simple alkyl groups (60). The results (Tables I and II) show a
clear superiority of the liquid crystal as a stationary phase
(columns A and B were of different film thickness). Whatever
the separative mechanisms may be it appears that ordered stationary phases are likely to demonstrate greater sensitivity to
the molecular shape of the principal solution conformers of a
diastereomeric pair (barring, obviously, additional interaction
potential such as hydrogen bonding or repellancy born of hydrophobicity, etc.) with the more linear molecule of the pair to
elute last (here, the transoid diastereomer). Figure 18 shows
the separation achieved for 4-methyl 3-heptanol carbamate
diastereomers.

TABLE I

GAS LIQUID CHROMATOGRAPHIC DATA FOR DIASTEREOMERIC AMIDES AND CARBAMATES.[a]

$$R^1R^2CHCNHCHCH_3C_6H_5$$

(with O double bond above C, and (R) below)

Compound; R^1, R^2	SE-54			Carbowax 20M			CpCC (Column A)		
	$k'(R,R)$	$k'(S,R)$	α	$k'(R,R)$	$k'(S,R)$	α	$k'(R,R)$	$k'(S,R)$	α
a. CH_3, C_2H_5	3.03	3.09	1.021	4.36	4.44	1.018	6.93	7.18	1.036
b. (a)-N-methyl	4.95	4.95	1.000				3.05	3.09	1.013
c. CH_3, $i-C_3H_7$	3.91	4.06	1.040	4.64	4.86	1.046	7.76	8.26	1.064
d. CH_3, $n-C_3H_7$	4.18	4.44	1.060	5.07	5.36	1.056	8.50	9.13	1.071
e. (d)-N-methyl	6.82	6.75	1.019				4.18	5.17	1.059
f. CH_3, $n-C_4H_9$	5.97	6.31	1.058	6.79	7.25	1.068	12.95	14.24	1.100
g. C_2H_5, $i-C_3H_7$	5.06	5.16	1.018	4.39	4.53	1.033	10.18	10.63	1.044
h. C_2H_5, $n-C_3H_7$	5.50	5.68	1.034	4.93	5.14	1.044	11.50	12.08	1.050
i. C_2H_5, $n-C_4H_9$	7.81	8.18	1.048	6.50	6.86	1.055	17.17	18.21	1.065
j. $n-C_4H_9$, $n-C_5H_{11}$							7.47	7.60	1.017
k. $n-C_5H_{11}$, $n-C_6H_{13}$	5.47	5.47	1.000				15.62	15.79	1.011
l. CH_3, C_6H_5	6.76	7.35	1.088	6.93	7.93	1.144	7.00	8.00	1.143
m. C_2H_5, C_6H_5	8.47	9.29	1.097	7.57	8.64	1.141	8.70	9.90	1.138

TABLE II

$$R^1R^2CHO\overset{O}{\overset{\|}{C}}NHCHCH_3C_6H_5$$
$$(\underline{R})$$

Compound; R^1, R^2	SE-54			Carbowax 20M			CpCC (Column B)		
	$k'(\underline{R},\underline{R})$	$k'(\underline{S},\underline{R})$	α	$k'(\underline{R},\underline{R})$	$k'(\underline{S},\underline{R})$	α	$k'(\underline{R},\underline{R})$	$k'(\underline{S},\underline{R})$	α
a. CH_3, $n\text{-}C_6H_{13}$	4.13	4.33	1.048	4.65	4.96	1.066	16.36	17.86	1.09
b. C_2H_5, $n\text{-}C_5H_{11}$	3.87	4.03	1.043	2.00	2.08	1.039	10.71	11.5	1.07
c. C_2H_5, $n\text{-}C_3H_7C\equiv C-$	4.63	4.80	1.036	8.92	9.23	1.037	10.21	10.57	1.04
d. C_2H_5, $n\text{-}C_3H_7CH=CH-(\underline{Z})$	3.60	3.73	1.037	4.12	4.38	1.065	6.28	6.64	1.06
e. C_2H_5, $n\text{-}C_3H_7CH=CH-(\underline{E})$	3.93	4.07	1.034	5.08	5.27	1.038	7.82	8.18	1.05
f. C_2H_5, $n\text{-}C_3H_7CH(CH_3)-$	(3.43, 3.50)			(3.31, 3.46, 3.58 [2])			(12.52, 12.99 / 13.20, 13.50)		1.038 / 1.017 / 1.027
g. CH_3, C_6H_5	4.43	4.57	1.030	7.96	8.31	1.044	20.10	21.8	1.08
h. CF_3, C_6H_5	2.73	2.67	1.025	4.31	4.31	1.000	2.89	8.43	1.07

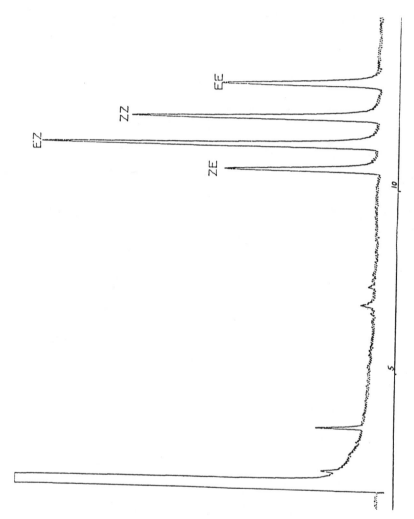

Figure 18. The 4 geometrical isomers of 11,13-hexadecadienal (Z,Z-isomer is the navel orangeworm pheromone). Conditions: 30M, LFV = 17 cm/s, 159°C, 50 = 1 split, 0.2 μl containing 0.2 μg material.

LITERATURE CITED

1. For a mellow introduction to this topic, see Ireland, R. E. Chapter 5, "Stereochemistry Raises its Ugly Head," in "Organic Synthesis, Prentice Hall Foundations of Modern Organic Chemistry Series," Reinhart, K. L., Jr., Ed. Prentice-Hall, Inc., Englewood Cliffs, 1969, pp. 100-122.
2. Izumi, Y.; Tai, A. "Stereodifferentiating Reactions," Academic Press, New York, 1977.
3. Valentine, D., Jr.; Scott, J. W. Synthesis 1978, 329-356.
4. Kagan, H. B.; Fiaud, J. C. "Topics in Stereochemistry," Vol. 10, John Wiley & Sons, New York, 1978.
5. ApSimon, J. W.; Seguin, R. P. Tetrahedron 1979, 35, 2797-2842.
6. Bartlett, P. A. Tetrahedron 1980, 36, 3-72.
7. Rossi, R. Synthesis 1978, 413-434.
8. Silverstein, R. M. "Enantiomeric Composition and Bio activity of Chiral Semiochemicals in Insects," in "Chemical Ecology: Odour Communication," Ritter, F. J., Ed., Elsevier/North-Holland Biomedical Press, New York, 1979, pp. 133-146.
9. Brand, J. M.; Young, J. Chr.; Silverstein, R. M. "Insect Pheromones: A Critical Review of Recent Advances in Their Chemistry, Biology and Application," in "Progress in the Chemistry of Natural Products," 37, Herz, W; Grisebach, H; Kirby, G. W. Springer-Verlag, New York, 1979.
10. Pirkle, W. H.; House, D. W. J. Org. Chem. 1979, 44, 1957-1960.
11. Kruse, K.; Francke, W; Konig, W. A. J. Chromatogr. 1979, 170, 423-429.
12. Schurig, V.; Koppenhoefer, B.; Buerkle, W. Angew. Chem. 1978, 90, 993-995.
13. Still, W. C. J. Am. Chem. Soc. 1979, 101, 2493-2495.
14. Martin, V. S.; Woodard, S. S.; Katsuki, T.; Yamada, U.; Ikeda, M.; Sharpless, K. B. Ibid. 1981, 103, 6237-6240.
15. Duhamel, L.; Plaquevent, J. -C. Ibid. 1978, 100, 7415-7416.
16. Valentine, D., Jr.; Blount, J. F.; Toth, K. J. Org. Chem. 1980, 45, 3691-3698.
17. Ojima, I.; Koguri, T.; Yoda, N. Ibid. 1980, 45, 4728-4739.
18. Caplar, V.; Comisso, G.; Sunjic, V. Synthesis 1981 2, 85-116.
19. Yamaguchi, S.; Mosher, H. S. Ibid. 1973, 38, 1870-1877.
20. Brinkmeyer, R. S.; Kapoor, V. M. J. Am. Chem. Soc. 1977, 99, 8339-8341.
21. Nishizawa, M.; Yamada, M.; Noyori, R. Tetrahedron Lett. 1981, 22, 247-250.
22. Midland, M. M.; McDowell, D. C.; Hatch, R. L.; Tramontano, A. J. Am. Chem. Soc. 1980, 102, 867-869.
23. Midland, M. M.; Tramontano, A. Tetrahedron Lett. 1981, 21, 3549-3552.

24. Midland, M. M.; Halterman, R. L. J. Org. Chem. 1981, 46, 12271229.
25. Rossiter, B. E.; Katsuki, T.; Sharpless, K. B. J. Am. Chem. Soc. 1981, 103, 464-465.
26. Enders, D.; Eichenauer, E. Angew. Chem. Int. Ed. (Engl.) 1979, 18, 397-399.
27. Ahlbrecht, H.; Bonnet, G.; Enders, D.; Zimmerman, G. Tetrahedron Lett. 1980, 21, 3175-3178.
28. Meyers, A. I.; Snyder, E. S.; Ackerman, J. J. H. J. Am. Chem. Soc. 1978, 100, 8186-8189, and references cited.
29. Davenport, K. G.; Eichenauer, H.; Enders, D.; Newcomb, M.; Bergbreiter, D. E. Ibid. 1979, 101, 5654-5659.
30. Meyers, A. I.; Williams, D. R.; Erickson, G. W.; White, S.; Druelinger, M. Ibid. 1981, 103, 3081-3087.
31. Meyers, A. I.; Williams, D. R.; White, S.; Erickson, G. W. Ibid. 1981, 103, 3088-3093.
32. Larcheveque, M.; Ignatova, E.; Cuvigny, T. Tetrahedron Lett. 1978, 19, 3961-3964, and J. Organomental. Chem. 1979, 177, 5-15.
33. Evans, D. A.; Takacs, J. M. Tetrahedron Lett. 1980, 21, 4233-4236.
34. Sonnet, P. E.; Heath, R. R. J. Org. Chem. 1980, 45, 3137-3139.
35. Meyers, A. I. Accts. Chem. Res. 1978, 11, 375-381.
36. Hashimoto, S.; Koga, K. Chem. Pharm. Bull. 1979, 27, (11), 2760-2766
37. Fraser, R. R.; Akiyama, F.; Banville, J. Tetrahedron Lett. 1979, (41), 3929-3932.
38. Frater, G. Helv. Chim. Acta 1979, 62, 2825-2832.
39. Schmierer, R.; Grotemeier, G.; Helmchen, G.; Selim, A. Angew. Chem. Int. Ed. Engl. 1981, 20, 207-208.
40. Schieh, H. M.; Prestwich, G. D. J. Org. Chem. 1981, 46, 4319-4321
41. Doolittle, R. E.; Tumlinson, J. H.; Proveaux, A. T.; Heath, R. R. J. Chem. Ecol. 1980, 6, 473-485.
42 Smith, L. R.; Williams, H. J. J. Chem. Ed. 1979, 56, 696-698.
43. Anderson, R. J.; Gieselmann, M. J.; Chinn, H. R.; Adams, K. G.; Henrick, C. A.; Rice, R. E.; Roelofs, W. L. J. Chem. Ecol. 1981, 7, 0000 and references cited.
44. Heath, R. R.; Doolittle, R. E.; Sonnet, P. E.; Tumlinson, J. H. J. Org. Chem. 1980, 45, 2910-2912.
45. Bergot, B. J.; Anderson, R. J.; Schooley, D. A.; Henrick, C. A. J. Chromatogr. 1978, 155, 97-105.
46. Still, W. C.; Mitra, A. J. Am. Chem. Soc. 1978, 100, 1927.
47. Korte, F. "Method Chimicum," Vol. 7B, Academic Press, New York, 1979, pp. 516-527.
48. Tamaki, Y.; Noguchi, H.; Sugie, H.; Sato, R.; Kariya, A. Appl. Entomol. Zool. 1978, 14, 101-113.
49. Suguro, T.; Mori, K. Agric. Biol. Chem. 1979, 43, 869-870.

50. Sonnet, P. E.; Heath, R. R. J. Chem. Ecol. (In press).
51. Guss, P. L.; Tumlinson, J. H.; Sonnet, P. E.; Proveaux, A. T. J. Chem. Ecol. (In press).
52. Dale, J. A.; Mosher, H. S. J. Am. Chem. Soc. 1973, 95, 512-519.
53. Guss, P. L.; McLaughlin, J.; Sonnet, P. E.; Tumlinson, J. H. (Manuscript in preparation).
54. Heath, R. R.; Jordan, J. R.; Sonnet, P. E.; Tumlinson, J. H. HRC and CC 1979, 12, 712-714.
55. Heath, R. R.; Jordan, J. R.; Sonnet, P. E. Ibid. 1981, 14, 328-332).
56. Helmchen, G.; Völter, H.; Schühle, W. Tetrahedron Lett. 1977, (16) 1417-1420.
57. Pirkle, W. H.; Hauske, J. R. J. Org. Chem. 1977, 42, 1839-1844.
58. Helmchen, G.; Nill, G.; Flockerzi, D.; Schühle, W.; Youssef, S. K. Angew. Chem. Int. Ed. Engl. 1979, 18, 62-63.
59. Bergot, B. J.; Baker, F. C.; Lee, E.; Schooley, D. A. J. Am. Chem. Soc. 1979, 101, 7432-7434.
60. Sonnet, P. E.; Heath, R. R. J. Chromatogr. (In press).

RECEIVED March 1, 1982.

Grandisol and Lineatin Enantiomers

FRANCIS X. WEBSTER and ROBERT M. SILVERSTEIN

State University of New York, College of Environmental Science and Forestry, Syracuse, NY 13210

The synthesis of grandisol(I) and lineatin(II) is discussed. Grandisol is a pheromone component in the cotton boll weevil and in Pissodes weevils. Lineatin is the pheromone of Trypodendron lineatum. The key intermediate, 3-methyl-5-oxo-3-cyclohexene-1-carboxylic acid(V), was resolved as the (-)-quinine salt. A new approach to determining the optical purity of carboxylic acids is discussed.

Grandisol (I) is a component of the aggregation pheromone of the cotton boll weevil, Anthonomus grandis (1), and is present in the hindgut of several species of male Pissodes weevils (2). Lineatin (II) is the aggregation pheromone of the female ambrosia beetle, Trypodendron lineatum (3,4). Since both compounds share the same carbon skeleton, synthesis schemes involving similar intermediates may be considered, and this paper describes several approaches to synthesizing the enantiomers of I and II. The importance of considering enantiomeric composition of chiral semiochemicals of insects has been discussed (5,6) and synthetic approaches have been reviewed (6,7).

Our concern with the grandisol enantiomers arises from our work with the aggregation pheromones of several species of Pissodes weevils (2). Although grandisol (I) is present (together with the corresponding aldehyde, grandisal) in the male hindgut and not in the female, the synthesized (racemic) compounds did not reproduce the attractivity of the males. We presume that it is necessary to reproduce the enantiomeric composition present in the males. However, the small specific rotation and the minute quantities available of grandisol precluded determination of the optical purity by optical rotation, and attempts to use a chiral shift reagent or a chiral derivate failed presumably because the functional groups is removed from the chiral centers.

The Hobbs and Magnus procedure (Scheme 1) (8) yielded only (+)-grandisol. Although Mori's procedure (Scheme 2) (9) yielded

0097-6156/82/0190-0087$06.00/0

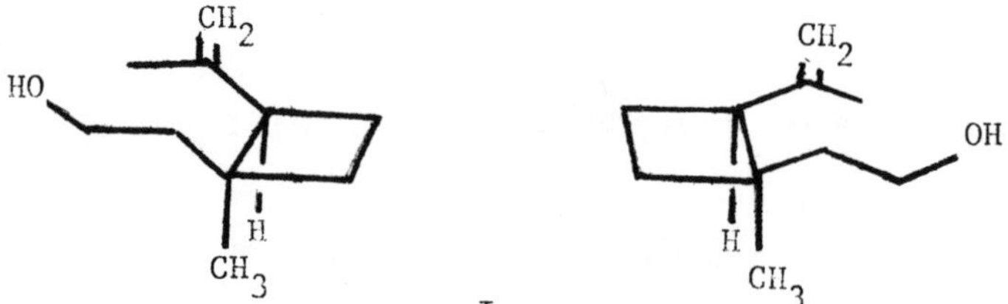
CH2
HO
H
CH3
CH2
OH
H
CH3

I

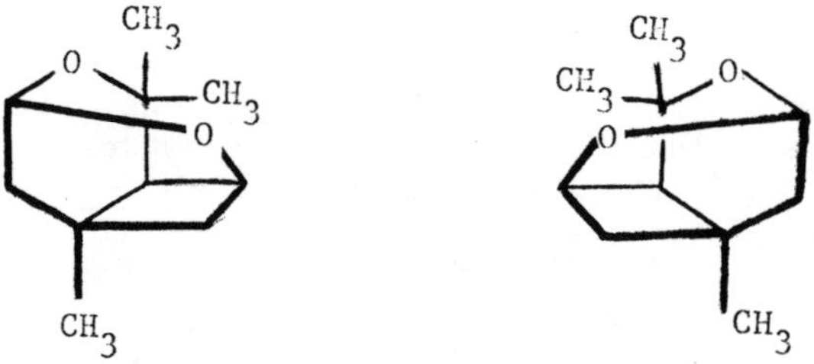
CH3
O
CH3
O
CH3
CH3
CH3
O
O
CH3

II

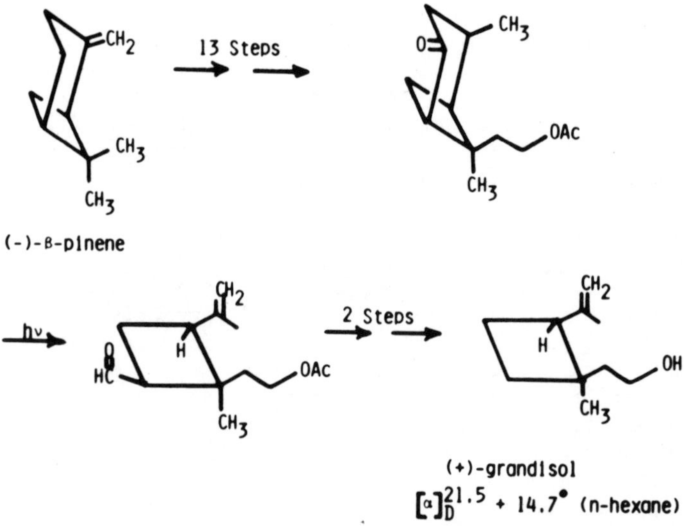
CH2
CH3
CH3
(-)-β-pinene
13 Steps
O
CH3
OAc
CH3
hν
CH2
H
HC
O
OAc
CH3
2 Steps
CH2
H
OH
CH3
(+)-grandisol
[α]D21.5 + 14.7° (n-hexane)

Scheme 1.

both enantiomers, neither enantiomer was better than 80%
optically pure, and the overall yield was poor. The second
synthesis by Mori and Tamada (Scheme 3) (10) resulted in an
optical purity of 94%, but only the (-)-enantiomer was obtained.
Mori and Tamada (10) reported that both enantiomers (mixed with
the other pheromone components) showed equal biological activity
for the boll weevil. In most cases, the antipode (an artifact)
of the naturally occurring enantiomer shows considerably less
activity (4). However, in most other chiral pheromone components,
the functional group is directly attached to the chiral center.
Possibly the greater distance between the functional group and
the chiral centers of grandisol explain the lack of discrimination, although the much greater distance in trogodermal (11) did
not interfere with discrimination by several species of Trogoder-
ma beetles.

The enantiomeric composition of lineatin in Trypodendron
lineatum was not determined, but the synthesized (racemic)
material was highly active in field tests (13,14). Mori et al.
(Scheme 4) (15) and Slessor et al. (Scheme 5) (16) synthesized
the lineatin enantiomers and reported similar optical purities,
but because different solvents were used, the optical rotations
cannot be compared. Since the (+) enantiomer was active in field
tests (16), we may assume that it is the naturally occurring
enantiomer; however the presence of the antipode cannot be ruled
out.

Exploratory Studies

Initially, our work resembled Mori's first synthesis of
grandisol (Scheme 2) (9). Whereas Mori partially resolved a keto
acid containing a bicyclo [3.2.0] heptane skeleton, we attempted
to resolve a keto acid containing a bicyclo [4.2.0] octane
skeleton (compound III). The synthesis of the acid was straight-
forward (Scheme 6): The cyanohydrin of cyclohexanone was
hydrolized and esterified, the α-hydroxy ester was smoothly
dehydrated with tosyl chloride in refluxing pyridine. Allylic
oxidation by chromic anhydride in acetic acid yielded the desired
enone ester in 55% yield. Although this yield could not be
improved by varying the conditions, the starting material could
be recovered for recycling. Photocyclization of the product with
ethylene proceeded in 98% yield. The bicyclo keto-ester was then
hydrolyzed to the desired keto acid, III.

The keto acid could not be resolved as easily as it was
synthesized. Formation of salts with various optically pure
bases invariably yielded oils. Ketalization of the keto ethyl
ester with ethylene glycol followed by reduction with lithium
aluminum hydride and treatment with Mosher's reagent, (+)-α-
methoxy-α-trifluoromethylphenylacetyl chloride (17,18,19), gave a
diastereomeric mixture (Scheme 7), which unfortunately was not
resolvable by HPLC.

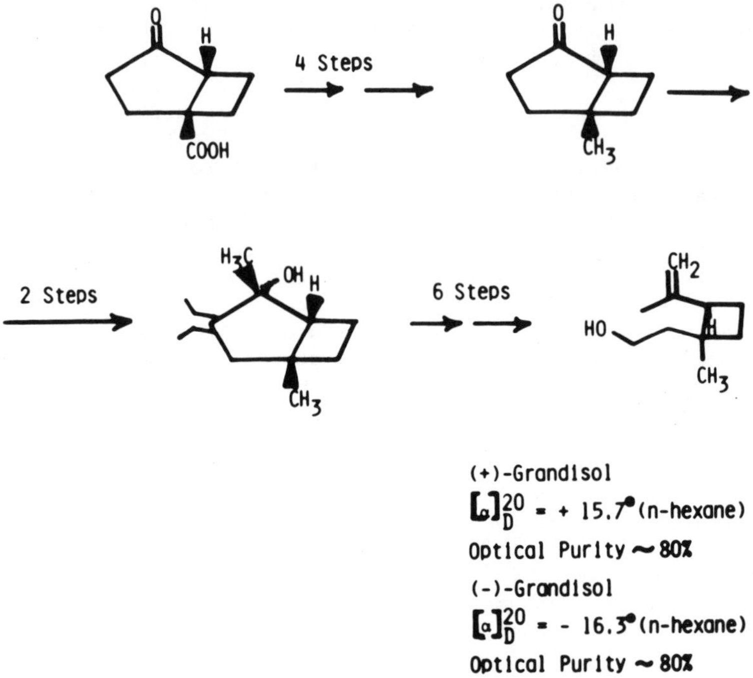
4 Steps
COOH
CH_3
2 Steps
H_3C
OH
6 Steps
HO
CH_2
CH_3
(+)-Grandisol
$[\alpha]_D^{20}$ = + 15.7° (n-hexane)
Optical Purity ~ 80%
(-)-Grandisol
$[\alpha]_D^{20}$ = - 16.3° (n-hexane)
Optical Purity ~ 80%

Scheme 2.

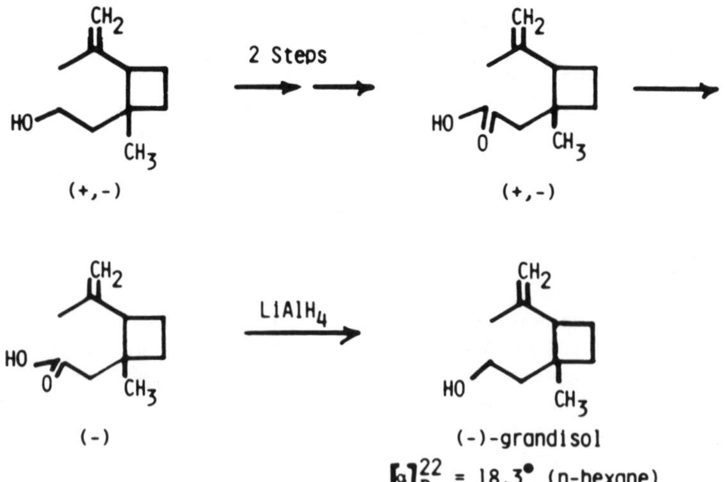
CH_2
2 Steps
HO
CH_3
(+,-)
(+,-)
$LiAlH_4$
(-)
(-)-grandisol
$[\alpha]_D^{22}$ = 18.3° (n-hexane)

Scheme 3.

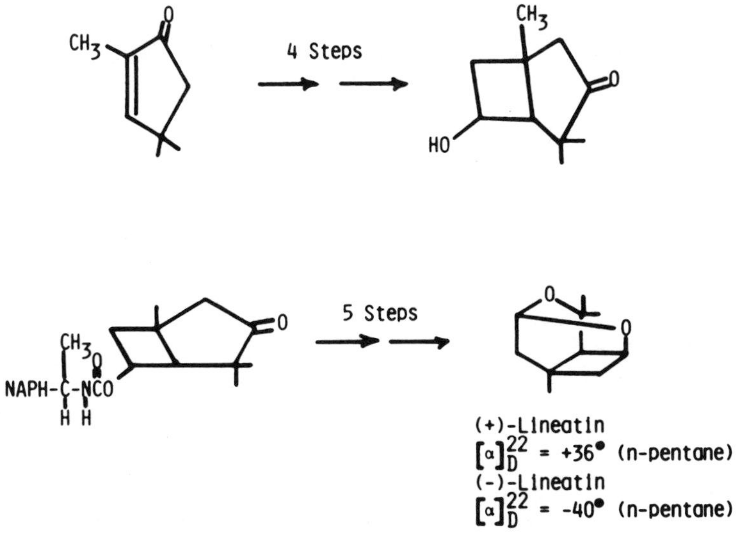

Scheme 4.

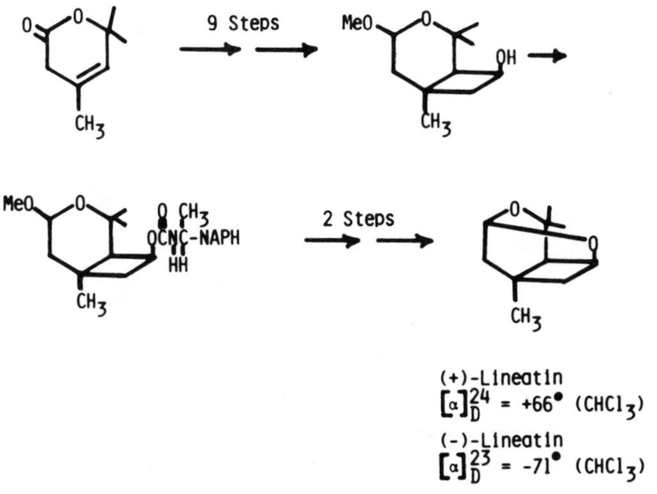

Scheme 5.

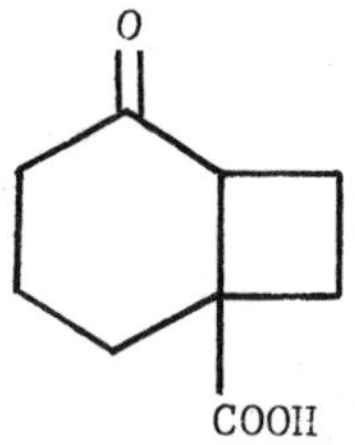

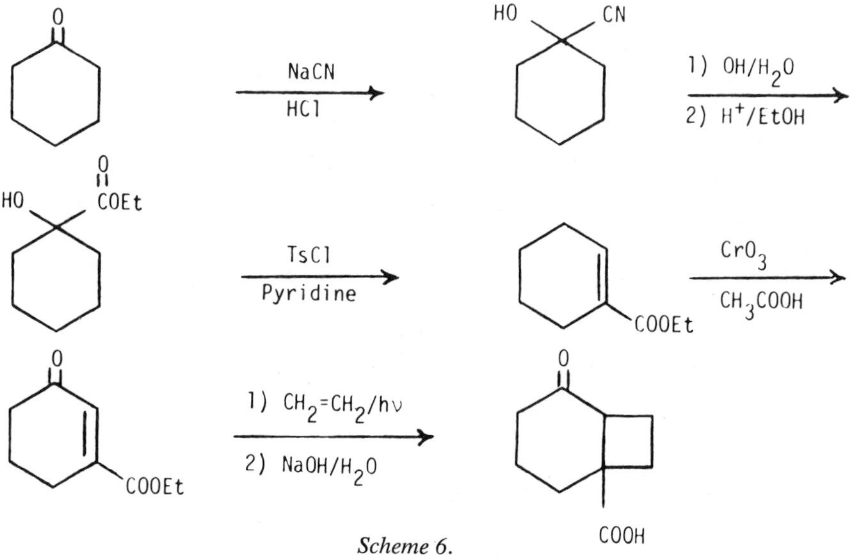

Scheme 6.

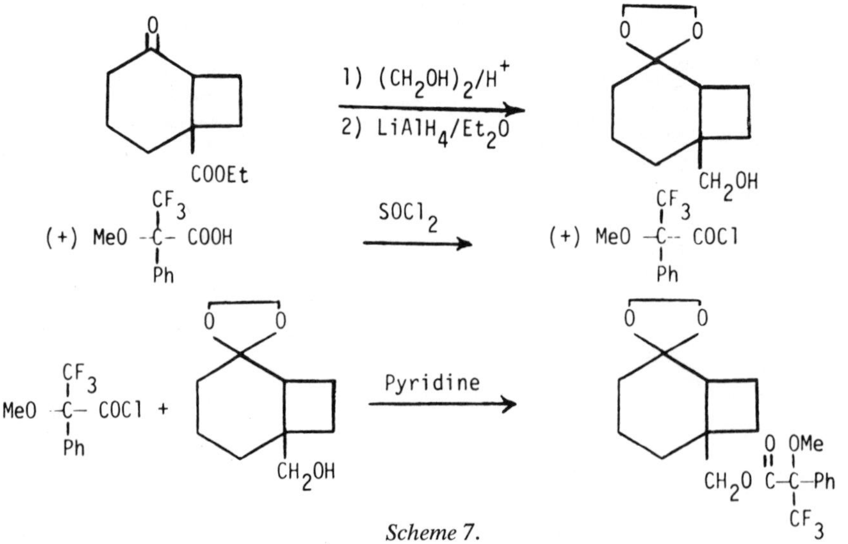

Scheme 7.

Since our alternatives at this point were of a less general nature, we modified our original goal of synthesizing the enantiomers of both grandisol and lineatin from a common precursor. Since we were involved with the white pine weevil, the synthesis of grandisol was given priority. Therefore, we cleaved the Mosher esters (MTPA esters), tosylated the neopentyl type alcohol (Scheme 8), and reductively cleaved the tosylate with lithium aluminum hydride in THF. The ketal moiety was removed in a two-phase system yielding the desired cis-bicyclic ketone. Zurflüh et al. (20) obtained the same compound as an intermediate in their stereospecific synthesis of racemic grandisol. This ketone was reduced at −78° C with lithium tri- tert-butoxyaluminohydride in THF yielding a single (probably cis) secondary alcohol. However, the diastereomeric MTPA esters could not be resolved by HPLC on a preparative scale even though the secondary OH group was directly attached to the chiral center.

Next, we investigated the synthesis of compound IV which might be used as a common intermediate in the synthesis of both pheromones. Although IV was never used for these purposes, our investigation led to some interesting results. Ethyl levulinate (Scheme 9) was brominated and dehydrobrominated in a one-pot sequence (21). The resulting enone was converted into the diene by treatment with trimethylsilyl iodide, formed in situ from trimethylsilyl chloride (22), in the presence of triethylamine. The diene was isolated by distillation in a non-aqueous work-up. The formation of a diene from an α,β-unsaturated ketone seems a general reaction. The diene was heated with maleic anhydride in toluene yielding a single adduct. This compound was brominated, the trimethylsilyl bromide formed was removed under vacuum, and dehydrobromination was carried out in pyridine. The anhydride was heated with water yielding the diacid; this compound spontaneously lost carbon dioxide forming the desired compound, IV. Note that this sequence yields only one isomer.

Since the excess trimethylsilyl bromide was difficult to remove, an alternative sequence was investigated (Scheme 10). After bromination of the silyl enol ether, the reaction mixture was poured into water to hydrolyze both the trimethylsilyl bromide and the anhydride. On heating this bromoacid as before, an unexpected compound was formed. This can be rationalized as follows: The reaction proceeds from the enol form, and the mechanism is formally 1,5 elimination of hydrogen bromide with concomitant loss of carbon dioxide. The second decarboxylation is analogous to the one seen earlier, and would be expected of the α,β-unsaturated ketone.

Synthesis and Resolution of the Common Key Intermediate

Keto acid V, a known compound, was used for the synthesis of both grandisol and lineatin. The first step of the two step synthesis is an ene reaction (23,24) of isobutylene with maleic

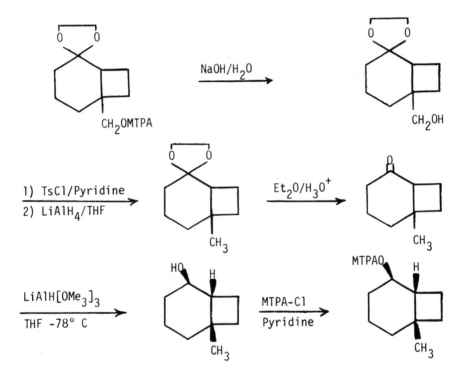

Scheme 8.

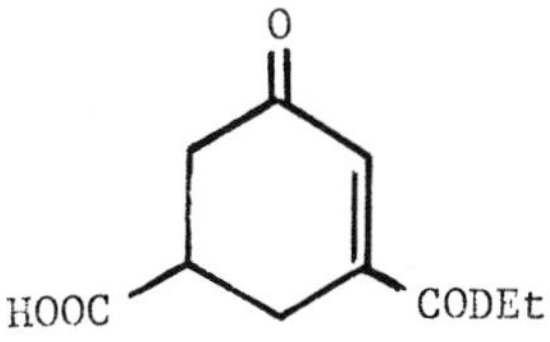
O
HOOC
CODEt

IV

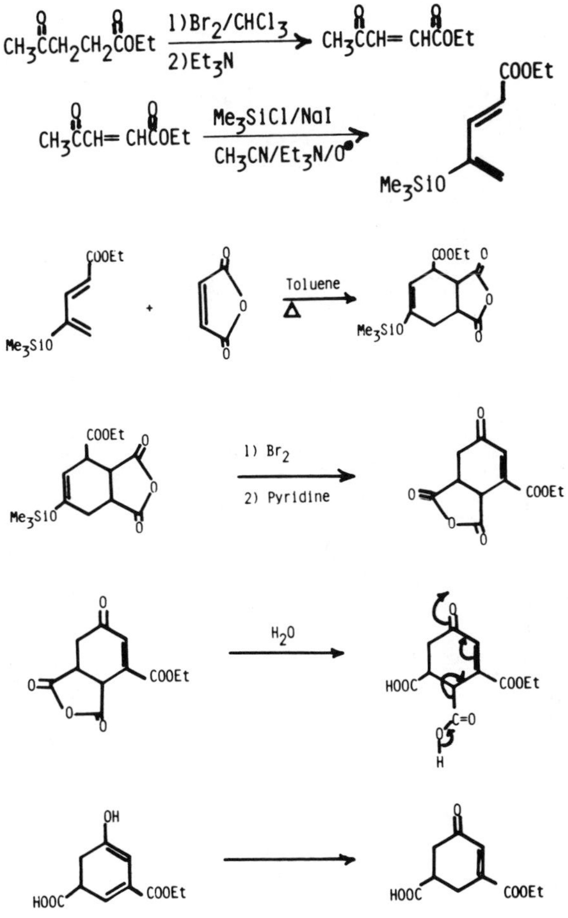
CH3CCH2CH2COEt
1)Br2/CHCl3
2)Et3N
CH3CCH= CHCOEt
Me3SiCl/NaI
CH3CN/Et3N/O°
COOEt
Me3SiO
Toluene
Δ
1) Br2
2) Pyridine
H2O
HOOC
C=O
OH

Scheme 9.

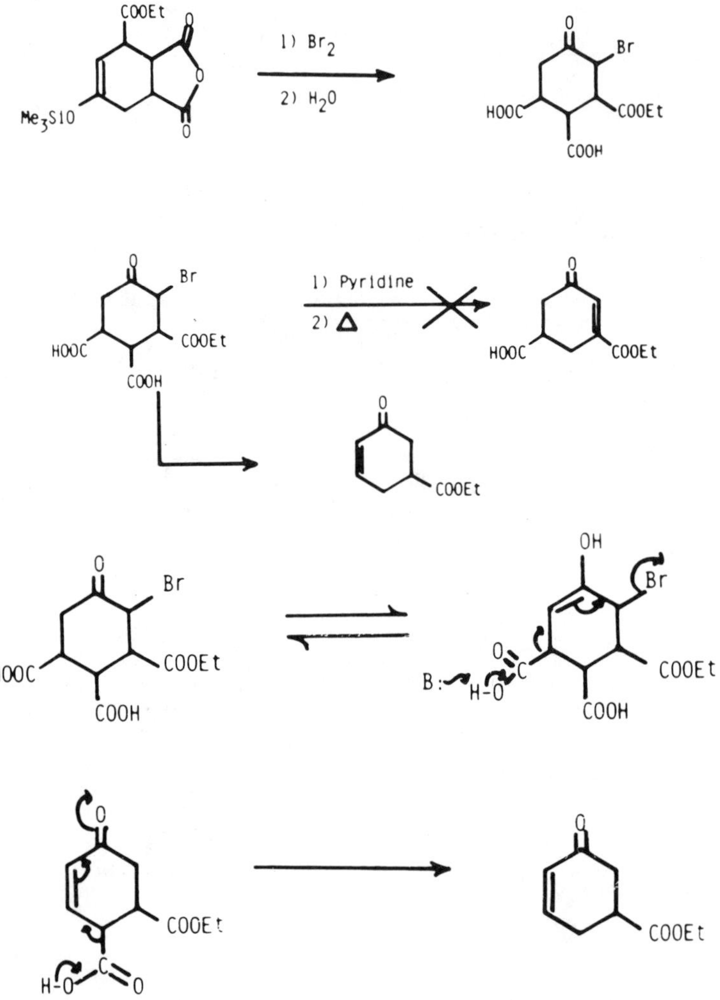
COOEt
O
O
O
Me3SiO
1) Br2
2) H2O
O
Br
HOOC
COOEt
COOH
O
Br
HOOC
COOEt
COOH
1) Pyridine
2) Δ
O
HOOC
COOEt
O
COOEt
O
Br
HOOC
COOEt
COOH
OH
Br
O
B:
H-O
C
COOEt
COOH
O
COOEt
H-O
C
O
O
COOEt

Scheme 10.

anhydride (Scheme 11). We were unable to repeat the reaction as reported (24) unless a small amount of hydroquinone was added. The product, methallylsuccinic anhydride, undergoes a clean rearrangement in hot polyphosphoric acid to yield enone acid V (25).

Racemic V could not be resolved with primary optically active bases presumably because they reacted in a Michael fashion. Formation of the quinine salt and five recrystallizations from a benzene/hexane mixture yielded an acid (after hydrolysis) whose specific rotation was $[\alpha]_D^{23}$ -47°. However, since the maximum specific rotation is unknown, an absolute method of determining optical purity is needed. The usual method is treatment with an optically pure derivatizing reagent to form covalent diastereomers, whose ratio can be determined by NMR or chromatography.

(-)-Quinine is an optically pure derivatizing reagent that forms a diastereomeric mixture of salts whose ^{13}C NMR spectrum would be expected to show individual peaks for the diastereomers present. A ^{13}C NMR of the salt formed from 1 equivalent of quinine and 1 equivalent of racemic acid (V) revealed two peaks for the carbon atom β to the carbonyl group. The NMR (25.2 MHz) of the salt of the partially resolved acid (Figure 1) indicates that the ratio of diastereomers is approximately three to one (optical purity ~75%). As is seen in the expansion, overlap of peaks precludes precise determination of enantiomeric composition, but a more powerful instrument should completely resolve the two peaks. We are currently investigating the generality of this procedure.

Synthesis of Grandisol

The first step in the conversion of enone acid V to grandisol is formation of the cyclobutane ring by irradiation of the acid in the presence of a continuous flow of ethylene (Scheme 12). Since the two chiral centers of grandisol are formed in this reaction, the product(s) of the reaction must be analyzed carefully. A priori, four isomeric products can be drawn: VI, VII, VIII, and IX. Since two of these products have a trans ring junction α to a ketone, they are unstable and can be epimerized to the more stable cis junction. Hence, a pair of isomers (VI and VII or VIII and IX) is acceptable because such a pair could be converted into a single isomer.

Analysis of the product revealed that two isomers were present. Figure 2 reproduces the ^{13}C NMR of the corresponding methyl esters (dimethyl sulfate/potassium carbonate/acetone). Since the ketone carbonyl (a) and adjacent bridge carbon (b) are each represented by two peaks, while the ester carbonyl (c) and the methoxy carbon (d) are each represented by a single peak, the mixture must consist of one of the acceptable pairs mentioned above.

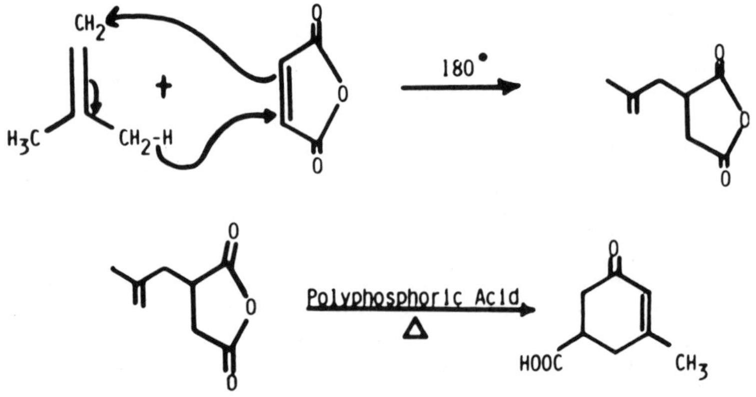

Scheme 11.

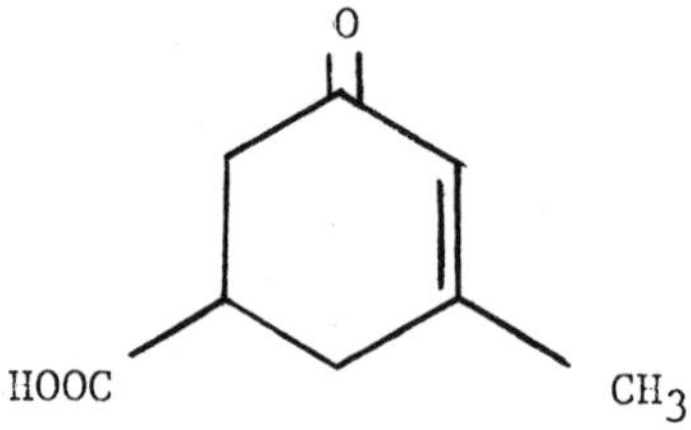

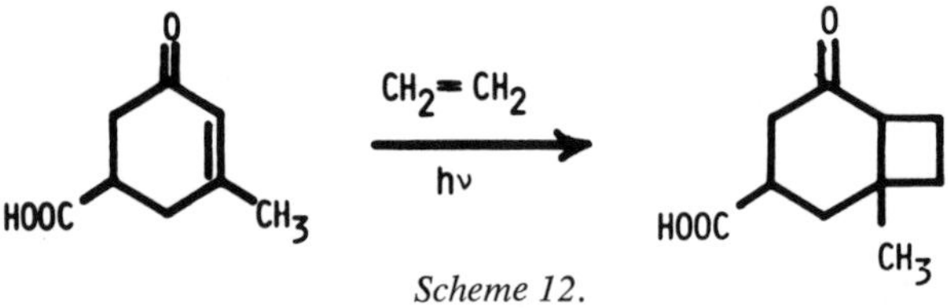

Scheme 12.

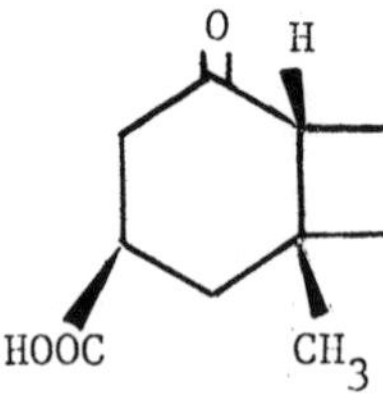

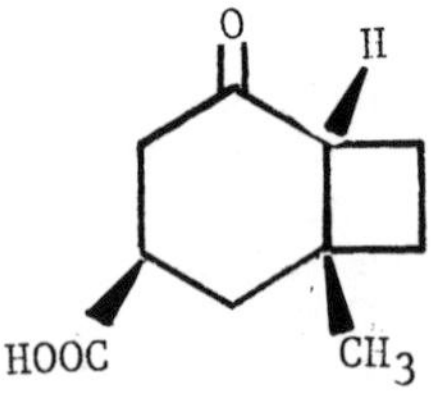

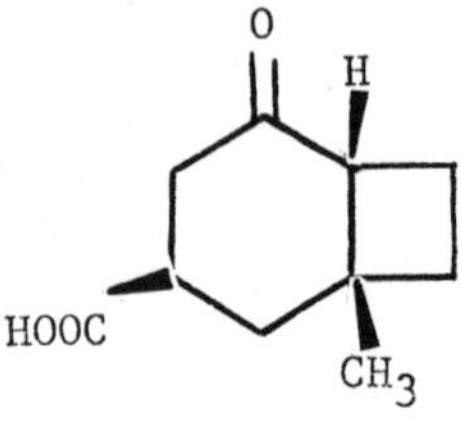

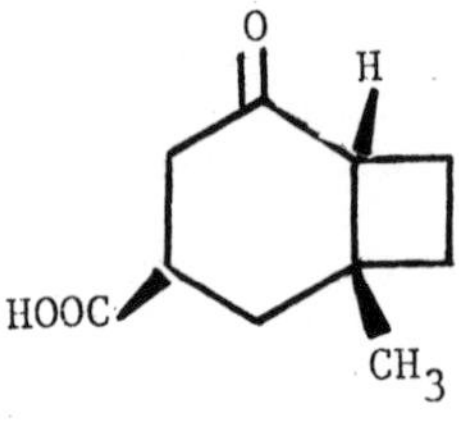

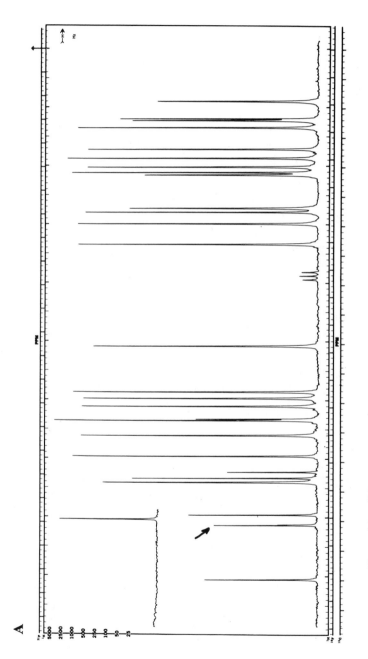

Figure 1A. ¹³C NMR spectrum (25.2 MHz) of the quinine salt of 3-methyl-5-oxo-3-cyclohexene-1-carboxylic acid in CDCl₃.

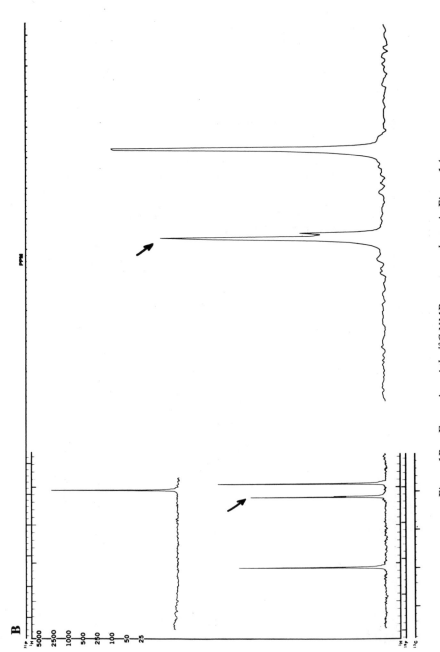

Figure 1B.   Expansion of the ^{13}C NMR spectrum shown in Figure 1A.

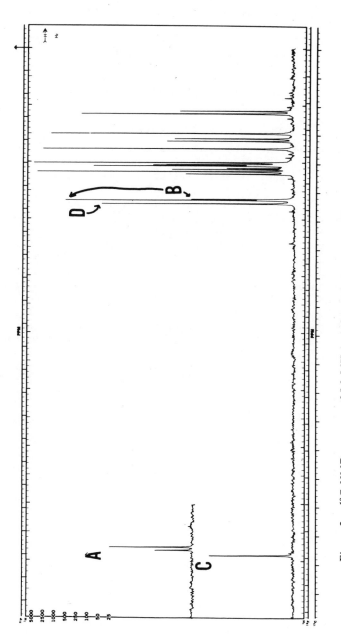

Figure 2. ¹³C NMR spectrum (25.2 MHz) of the methyl ester of acid V in CDCl₃. Key: A, ketone carbonyl; B, adjacent bridge carbon; C, ester carbonyl; D, methoxy carbon.

The rest of the synthesis (Scheme 13) is completely stereo-specific and most of the steps are known (20). The bicyclic acid was oxidatively decarboxylated with lead tetraacetate and copper acetate (21). The resulting enone was alkylated with methyllithium giving a single crystalline allylic tertiary alcohol. This compound was cleaved with osmium tetroxide and sodium periodate. Inverse addition of the Wittig reagent effected methylenation in 85% yield. Finally, the acid was reduced with lithium aluminum hydride to grandisol.

Synthesis of Lineatin

Scheme 14 outlines the synthesis that we have developed for lineatin. Photocyclization with allene gave predominantly one isomer (26,27), which, after purification, was obtained as a single crystalline compound. This reaction is very satisfactory for chiral induction. The next step in the synthesis is the oxidative decarboxylation seen earlier in the grandisol synthesis. Oxidative cleavage with osmium tetroxide and sodium periodate yielded the trifunctional compound, which was converted into the corresponding acetal ketal ester. This ester was alkylated with methyllithium and the ketal group was hydrolyzed. Although the following reduction shown in the scheme has not yet been worked out, the plan is designed to save one step. If the ketone can be reduced kinetically while the hemiacetal remains intact, then the synthesis is virtually complete. However, if the reduction is unsuccessful, we shall oxydize the hemiacetal and reduce both groups with DIBAH. In either case, 4,6,6-lineatin will be formed by closing the third ring with acid catalysis.

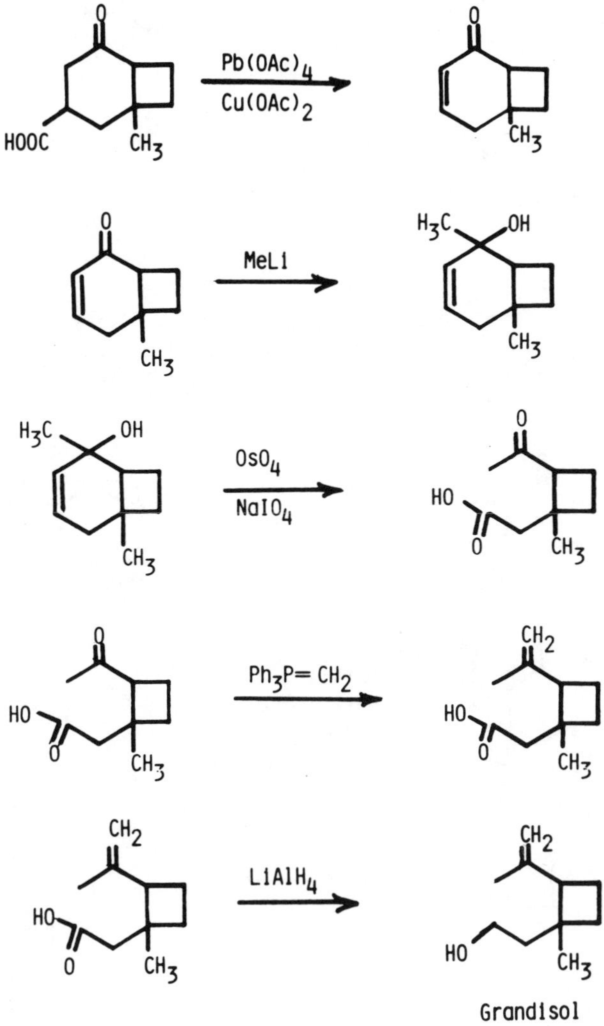
O
HOOC
CH3
Pb(OAc)4
Cu(OAc)2
MeLi
H3C
OH
OsO4
NaIO4
HO
Ph3P= CH2
CH2
LiAlH4
Grandisol

Scheme 13.

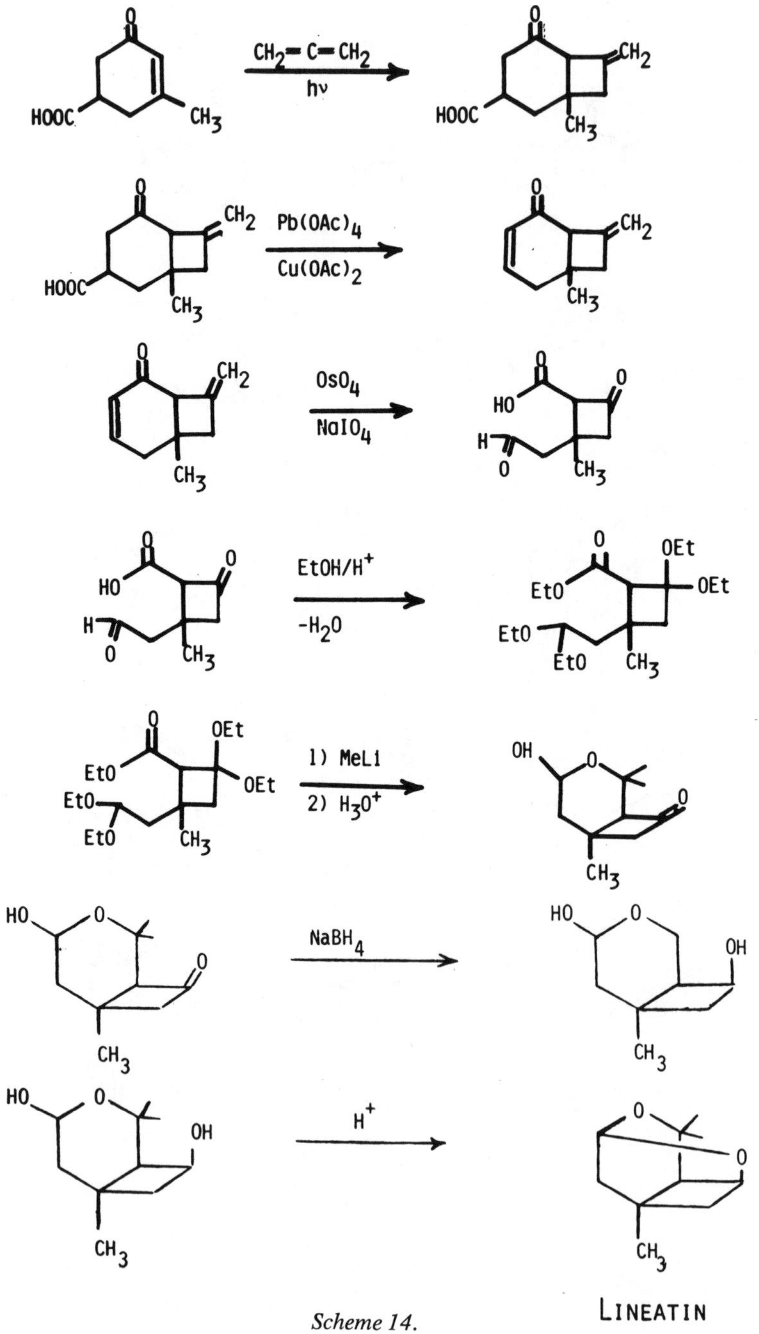

HOOC
CH_3
$CH_2=C=CH_2$
hν
CH_2
$Pb(OAc)_4$
$Cu(OAc)_2$
OsO_4
$NaIO_4$
HO
H
EtOH/H+
-H_2O
EtO
OEt
1) MeLi
2) H_3O^+
OH
$NaBH_4$
H+
LINEATIN

Scheme 14.

Literature Cited

1. Hedin, P.A.; Gueldner, R.C.; Thompson, A.C. In "Pest Management with Insect Sex Attractants"; Beroza, M. Ed.; American Chemical Society: Washington, 1976; pp. 30-52.
2. Booth, D.C.; Phillips, T.W.; Claesson, A.; Silverstein, R.M.; Lanier, G.N.; West, J.R. J. Chem. Ecol., in press.
3. Borden, J.H.; Handley, J.R.; Johnston, B.D.; MacConnell, J.G.; Silverstein, R.M.; Slessor, K.N.; Swigar, A.A.; Wong, D.T.W. J. Chem. Ecol. 1979, 5, 681-689.
4. MacConnell, J.G.; Borden, J.H.; Silverstein, R.M.; Stokking, E. J. Chem. Ecol. 1977, 3, 549-561.
5. Silverstein, R.M. In "Chemical Ecology: Odour Communication in Animals"; Ritter, F.J. Ed.; Elsevier/North Holland: Amsterdam, 1979; 133-158.
6. Brand, J.M.; Young, J.C.; Silverstein, R.M. In "Progress in the Chemistry of Organic Natural Products"; Herz, W.; Griseback, H.; Kirby, G.W. Eds.; Springer-Verlag: Vienna, 1979; 1-190.
7. Mori, K. In "The Total Synthesis of Natural Products", Vol. 4; ApSimon, J. Ed.; Wiley: New York, 1981; 1-183.
8. Hobbs, P.D.; Magnus, P.D. J. Am. Chem. Soc. 1976, 98, 4594-4600.
9. Mori, K. Tetrahedron 1978, 14, 915-920.
10. Mori, K.; Tamada, S. Naturwissenschaften 1978, 65, 653-654.
11. Silverstein, R.M.; Cassidy, R.F.; Burkholder, W.E.; Shapas, T.J.; Levinson, H.Z.; Levinson, A.R.; Mori, K. J. Chem. Ecol. 1980, 6, 911-917.
12. Borden, J.H.; Handley, J.R.; Johnson, B.D.; MacConnell, J.G.; Silverstein, R.M.; Slessor, K.N.; Swigar, A.A.; Wong, D.T.W. J. Chem. Ecol. 1979, 5, 681-689.
13. Borden, J.H.; Oehlschlager, A.C.; Slessor, K.N.; Chong, L.; Pierce, H.D., Jr. Can. Entomol. 1980, 112, 107-109.
14. Mori, K.; Sasaki, M. Tetrahedron 1980, 36, 2197-2208.
15. Klimetzek, D.; Vité, J.P.; Mori, K.Z. Angew. Entomol. 1980, 89, 57-63.
16. Slessor, K.N.; Oelschlager, A.C.; Johnston, B.D.; Pierce, H.D., Jr.; Grewal, S.K.; Wickremesinghe, L.K.G. J. Org. Chem. 1980, 45, 2290-2297.
17. Koreeda, M.; Weiss, G.; Nakanishi, K. J. Am. Chem. Soc. 1973, 95, 239-240.
18. Weiss, G.; Koreeda, M; Nakanishi, K. J. Chem. Soc. Chem. Comm. 1973, 565-566.
19. Dale, J.A.; Dull, D.L.; Mosher, H.S. J. Org. Chem. 1969, 34, 2543-2549.
20. Zurflüh, R.; Durham, L.L.; Spain, V.L.; Siddall, J.B. J. Am. Chem. Soc. 1970, 92, 425-427.
21. McMurry, J.E.; Blaszczak, L.C. J. Org. Chem. 1974, 39, 2217-2222.

22. Olah, G.A.,; Narang, S.C.; Gupta, B.G.B.; Malhorts, R.
     J. Org. Chem. 1979, 44, 1247-2451.
23. Alder, K.; Pascher, F.; Schmitz, A. Ber. 1943, 76B, 47-49.
24. Phillips, D.D.; Johnson, A.W. J. Am. Chem. Soc. 1955, 77,
     5977-5981.
25. Noyce, D.S.; Dolby, L.J. J. Org. Chem. 1961, 26, 1732-1737.
26. Guthrie, R.W.; Valenta, Z.; Wusner, K. Tetrahedron Lett.
     1966, 4645-4654.
27. Kelly, R.B.; Zamecnik, J.; Beckett, B.A. Can. J. Chem.
     1972, 50, 3455-3464.

RECEIVED February 24, 1982.

Recent Pheromone Research in the Netherlands on Muskrats and Some Insects

F. J. RITTER, I. E. M. BRÜGGEMANN, J. GUT, and C. J. PERSOONS

Netherlands Organization for Applied Scientific Research TNO, Delft, Netherlands

The paper describes research on three pests, introduced from America into Europe: the muskrat, Ondatra
zibethicus, the American cockroach, Periplaneta
americana, and the beet armyworm, Spodoptera exigua.
Screening of a variety of muskrat lures in the field
and in pens showed that extracts of the preputial
glands of male muskrats were most attractive. GC/MS
analysis proved that the composition of the musk in
these glands changes with season. The extracts contain at least 13 macrocyclic ketones, the major ones
of which are the (Z)-5 isomers of cycloheptadecenone
and cyclopentadecenone, and the corresponding saturated ketones. They are probably biogenetically
related to the fatty acids found to be present in
the same glands.
The identification of periplanone-A, one of the two
sex pheromones of the American cockroach, is reported. It is a sesquiterpenoid, $C_{15}H_{20}O_2$, which readily
isomerizes to a more stable but biologically inactive compound. The structures of both compounds are
deduced from spectroscopic data. Periplanone-A is
7-methylene-4-isopropyl-12-oxa-tricyclo $[4.4.2.0^{1,5}]$
-9-dodecene-2-one:

From females of the beet armyworm, Spodoptera exigua,
the (Z,E)- and (Z,Z)-tetradecadienyl acetates, the
(Z)-9- and (Z)-11-tetradecenyl acetates and tetra-
decanyl acetate were isolated and identified. Synthetic mixtures of some of these components attract
male moths in field tests.

Mammalian Pheromones: An Area Neglected by Applied Biologists and Pesticide Chemists

Legends and old historical records prove that even in ancient
times man has shown interest in odourous secretions of animals,
such as the musk deer and the civet, to use in precious perfumes.
This is the main reason why chemists have analysed, synthesized
and mimicked the components of these secretions long before any
insect pheromone was isolated and identified.

One may therefore wonder why so little attention has been
paid by scientists to the behavioural role, the chemistry and the
application of such mammalian chemical signals.

This partly can be attributed to great experimental difficulties in studying mammalian pheromones (1). Compared with behavioural and physiological bioassays used in insect pheromone
research, reliable, reproducible and significant bioassays for
mammalian pheromones are hard to device. Such assays are essential for chemists to guide their isolation and identification and
for biologists to determine the biological significance of a compound for the behaviour of the animal.

The reaction of mammals to olfactory stimuli is certainly
much more influenced by experience, environmental factors and
other sensory inputs than is the reaction of insects to pheromones. Moreover, the same odour may serve different behavioural
functions and these may be dependent on such variables as season,
age, etc.. Even the word "pheromone" for chemical signals in
mammals has been questioned (2). Although the objections were
rebutted (3) the term is today still a matter of controversy or
at least discussion (4).

Another reason for the relative lack of interest in mammalian
pheromones lies in the fact that - apart from the use in perfumes - only one actual application of mammalian pheromones is
known so far. This is not in connection with a pest but with a
very useful animal, the domestic pig (5). The saliva and sweat
glands of the sexually aroused boar contain two steroids with a
musky smell, Δ^{16}-androstenol and Δ^{16}-androstenone. These are
actual pheromones emitted by the male and eliciting a characteristic behaviour in the female. Sows in heat react to the scent of
these compounds by assuming a characteristic copulating stance.
The pheromone is commercially available as an aid in artificial
insemination, to show when a young sow is in oestrus.

The major part of mammalian pheromone research, much of
which is mentioned in recent books and symposium proceedings
(5-10) is not directed to any application. Trappers, however,
have used mammalian secretions in their lures long before the

term pheromone was coined, and the possibility of application of
mammalian chemical signals has been suggested by scientists like
Christiansen and Døving for control of small rodents (11) and
by Marsh and Howard for control of rodents in general (12).
Although they positively evaluate the potentialities of these
agents in rodent control, the few publications in this area relate
rather to laboratory experiments than to actual application in
pest control.

This is particularly surprising in the case of common rats
since we know that the male preputial gland contains a mixture of
aliphatic acetates which appears to act as an attractive sex
pheromone to the female (13). This gland also appears to have an
important function in conspecific chemical communication in mice.
Its weight is inversely related to population density (14) and is
larger in dominant than in subordinate males (15). Extracts of
these glands attract female mice (16) and preputialectomy affects
male fighting behaviour (17).

Many other skin glands of mammals also produce odourous
secretions to which behavioural functions are attributed: marking
of territories, recognition of species, of sex or individuals,
etc.. They may also serve in familiarization with strange sur-
roundings e.g. "gaining confidence" as exemplified by the behavior of rabbits (1). A recent review on this subject is given
by Müller-Schwarze (18).

The Muskrat: A Popular Fur Animal, A Nightmare in The Netherlands

In most countries in the Northern hemisphere the muskrat,
Ondatra zibethicus, is mainly appreciated for its beautiful muskrat or "bisam" fur. It was for this reason that it was introduced
from America into Czechoslovakia and other European countries
early in the 20th century. In the Netherlands, however, it is now
a major pest, damaging the banks of the many canals and dikes,
so vital for a country lying largely below sea level. This threat
is evident from the schematic drawing of Figure 1.

In the beginning of World War II the rats invaded the Netherlands, which for brevity, I will call Holland, although this is a
part of the Netherlands in which this rodent is still rare.
Unlike the German army, however, the muskrat first crossed the
Belgian, and later the German border. Between 1950 and the present
day the muskrat has constantly expanded its territory and it now
occupies the larger part of the country, in spite of constant
trapping by government-employed professional trappers.

Therefore the Dutch government decided to look for alternative methods to stop and control the intruder, and our research
group was asked to look into the possibilities of using pheromones or any other semiochemicals.

There is a vast literature about the muskrat. A review by
Hoffmann (19) mentions 6858 references, 1669 of which came from
the USSR. Not all of these sources are readily available, but from
translated titles and from the English and German literature it

seems that little has been published about chemical communication
in muskrats.

Ill-defined concoctions have for centuries been used to lure
muskrats in the USA, but we have looked in vain for published data
which prove the effectiveness of these lures. Only one very brief
statement by Williams (20) about the effectiveness of addition of
"muskrat scent" to carrot-baited traps was found in the Journal
of Wildlife Management of 1951. A table showed that the "average
trapping success" increased from 23.6 to 42.8%, but the two series
of experiments (controls and scented carrots) were not made simul-
taneously, so these results hardly prove the attractancy of the
scent.

One of our first targets was therefore to find out whether
or not attraction by muskrat musk or any other traditional lure
can be demonstrated. This was initially investigated by applying
trapping methods normally used by trappers in Holland; this was
done in collaboration with experienced trappers and using a
variety of lures.

Methods for Field-Trapping Muskrats

In Holland the muskrat usually lives in burrows in dams,
dikes and banks of the canals and lakes, as depicted in Figure 1.
The "muskrat houses" or "winterhuts", well-known in the USA, are
relatively rare in Holland. They are usually found in wetlands
and marshes where no dikes are available. They are conspicuous
and relatively easy to find. This is not the case with the burrows in the banks and dikes. The entrances are below water level
and can only be observed when the level drops. Experienced trappers have, however, a variety of clues to detect the presence of
an entrance, e.g. by swimming trails, or heaps of sand removed
from the burrows.

In Holland, the use of poisoned baits which could kill other
wildlife, and ground traps which would not kill the animal immediately, are prohibited. The devices most often used are:
- baited traps, afloat on rafts, in which apples are often used
  as bait,
- "Conibear traps", placed at the entrance below the water surface,
- metal hoop-nets, either placed at the entrance of a burrow or
  at the interconnection of waterways or under a bridge or,
  according to a recently developed technique, in "fake entrances",
  made of plastic tubes or pipes (Figure 2),
- a variety of floating "valve-traps" or other contraptions,
  usually baited with apples.

A number of field tests using such methods were made in order
to screen a variety of lures, either obtained through the help of
Prof. Müller-Schwarze of New York State University in Syracuse,
with whom we collaborate, or made by ourselves from collected
glands or commercially available fragrances.

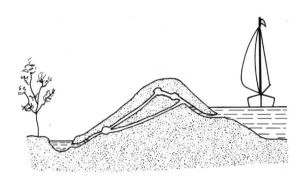

Figure 1. *Schematic representation of the system of muskrat burrows and nests in a dike of a characteristic Dutch canal.*

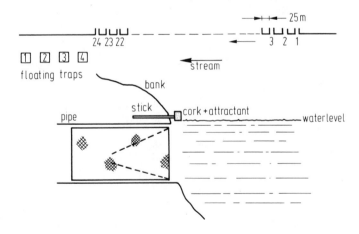

Figure 2. *Top: Scheme of 4 rafts with baited traps and 24 funnel traps placed in "artificial burrows" at the Berkel on the German border. Bottom: Funnel trap in plastic pipe placed below the water surface. A scented cork is stuck just above the trap. Results of 2 series of experiments with such traps are shown in Figures 3 and 4.*

Results of Field Tests with Muskrat Lures

Field tests were made at a variety of locations near the German or Belgian border and in an area between rivers, south of the Rhine.

In this paper we cannot sum up all the results, many of which were inconclusive. We attribute the latter to an insufficient number of muskrats in the test area (apart from trapping, no reliable method was available to estimate the size of the population) or to the lack of attractiveness of the lures in mid-summer, outside the mating or migration season. We have to stress, therefore, that the following results must be considered as preliminary ones and are a selection of only those results which seem to be statistically significant.

The most convincing evidence of the effectiveness of some lures was obtained at the Berkel, a little border river between Germany and Holland. Here we mainly used the method with artificial entrances, in which funnel-shaped traps were placed (Figure 2). In the slope of the bank 24 holes were dug in which pipes with traps were placed. Three different lures were tested in sixfold against an equal number of controls. In addition, 4 floating traps, baited with apples, and the same lures were tested. Five drops of lure were applied to corks placed on sticks above the funnel traps and similar corks were used on the rafts. At least once a week the traps were checked and fresh drops of lures applied. The results with the funnels are shown in Figure 3. The samples tested were:

A. Scent gland extracts of male muskrats, caught in March in the same vicinity but somewhat more to the North (the province of Drenthe). The two batches tested (26 and 29 in Figure 3) were obtained by centrifugation of homogenized glands in ethanol. The supernatant used represented 5 ml extract per gland, and the 5 drops corresponded roughly with 0.1 ml or 1/50 of the extract from one gland.
B. Commercially available American "Muskrat Musk" (Perkins Hill Ltd., Cazenovia, N.Y.) (sample 28 in Figure 3).
C. A mixture of oil of sweet flag in glycerine (1:2), made according to one of the traditional trappers recipes, kindly provided by Prof. M.M. Alexander, Syracuse, N.Y. (sample 13 in Figure 3).

As the Figure shows, only 4 muskrats were caught during the whole test period - which ran from September 15, 1980 to June 5, 1981 - in the controls compared with 19 in the traps with the home-made gland extracts. In the first two months these numbers were even 1 and 10. With sweet flag and American musk the total catches over the whole period were 6 and 9, respectively. Unfortunately, the water level dropped considerably during three periods, as indicated in the Figure, but this does not affect the conclusion that the gland extract was best. In total 4 females and 34 males were caught.

The baited traps on the rafts seemed less suited for screening attractants, as the apples attract hungry muskrats and only slight, if any, improvement was obtained by adding the lures, with the exception of the American musk which seemed to enhance the attractivity of the baits during the months March and April. An additional problem with the baited traps is that they also catch other animals. In the test period 37 male and 2 female muskrats were caught this way, but also 9 common brown rats. At the Hegebeek, a nearby location, at the slopes of a small pool near a lock, the lures A, B and C could only be tested singly, interspaced by two non-scented traps (blanks) (Figure 4). Here again, the extracts of the Dutch muskrats came out best, although the other traps and even one of the blanks were not ineffective.

According to the trappers, the predominance of males is not unusual at these locations near the German border. It is attributed to the observation that during migration the males travel first.

Field tests with baited traps and the same three lures in the middle of the country between the branches of the Rhine, did not allow conclusions about the attractancy of the lures to be reached, but they showed that here about equal numbers of males and females were caught. At this location a stabilized colony of muskrats had settled.

These preliminary results seem to confirm that extracts of muskrat preputial glands and possibly also of sweet flag can indeed enhance the catches of muskrats, especially in scented funnel traps in artificial pipes at river banks, where migrating muskrats pass. The animals caught are then mainly males which preferentially enter artificial pipes scented with extracts of musk glands which - it should be remarked here - are produced by other males.

Screening Lures with Muskrats in Captivity

A number of muskrats caught in live-traps near the Belgian border were used for screening lures in two series of tests.

The first one was made in a concrete basin, 7 m long, 1.5 m wide and 1.6 m deep (Figure 5), in which 4 to 8 muskrats were kept. Nest boxes were made in the corners and two counting cages were placed on a raft near the middle of the basin. Animals visiting the cages were counted mechanically by a device attached to a flap which could swing in both directions.

A variety of lures was screened here in dual choice tests. A perforated aluminum capsule, containing five drops of the test sample on filter paper, was attached to the back wall of one cage; the other cage carried a blank capsule. The average test period was one week, in the middle of which the places of the cages were interchanged and the samples were renewed. After each test the cages were thoroughly rinsed with clean water. The counts were recorded daily and the tests were made at least in duplicate.

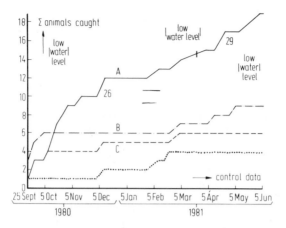

Figure 3. Results of the screening experiment at the Berkel on the German border. Key: A, (———), Dutch muskrat musk 26 and 29; B, (– · – ·), American muskrat musk 28; C, (– – –), sweet flag 13; and blank (· · ·).

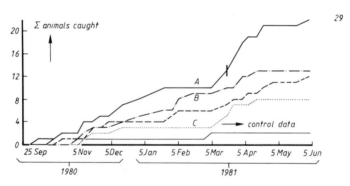

Figure 4. Results of the screening experiment at the Hegebeek on the German border. Key: A, (·), Dutch muskrat musk 26 and 29; B, (– · – ·), American muskrat musk 28; blank 1, (– – –); C, (· · ·), sweet flag 13; and blank 2, (———).

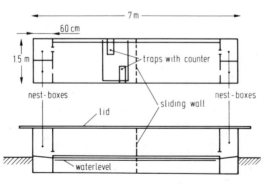

Figure 5. Schematic representation of the basin for keeping muskrats at Rijswijk. Top shows view from above, and bottom shows view from the side.

Within the framework of this paper it is not possible to discuss all the screening results. For those who are interested, a detailed report (21) (in Dutch) is available. It may suffice here to relate that the home made gland extract A (batch 29, which also performed best in the field tests of Figures 4 and 5) was the best attractant, followed by a mixture of the oils of sweet flag and catnip (recipe obtained through the courtesy of Prof. Alexander) and the American muskrat musk, which was also used in the field tests. As an example, the results of one of these tests is shown in Figure 6.

The second series of tests is at present being carried out in an "olfactometer", which consists of a round "muskrat pen" with a diameter of 10 m, containing a swimming pool (Figure 7). Eight cages which are connected to electric counting devices and a recorder, are equally spaced round the periphery of the pen (Figure 8).

At present the behaviour of the muskrat colony in the pen and the individual reactions of the muskrats to different lures is being studied by students of Prof. Wiepkema in Wageningen. They watch the animals from an observation hide during the dark and light periods of greatest activity and keep record of the number and duration of the visits to scented cages and controls. This pen and its equipment have only recently been installed and are being tested out at present. We hope that they will contribute to a better understanding of the general behaviour and the chemical communication of muskrats and at the same time will allow us to screen different lures in a systematic manner, to determine their optimal concentrations, and to study the effect of the various components found in the extracts of the scent gland.

Parallel studies on the behavioural function of the musk odour of muskrats are being done in the USA by Prof. Müller-Schwarze in Syracuse, N.Y.

Possible Function of Musk Gland Secretion and Seasonal Changes of its Chemical Composition

Function. The preputial gland of the muskrat enlarges greatly in the breeding season. It is reasonable to assume that this strong-smelling secretion has at least one and possibly several functions in the chemical communication between members of this species.

In muskrats these glands are present in both sexes, but they are much larger in males, in which size of the glands and strength of the odour secretion parallel the cyclic changes in testes size.

The variety of possible functions of preputional gland secretions of mammals have been mentioned before. It is still not clear which of these functions apply to muskrats. They may include sex attraction, repulsion of members of the same sex, marking of nests or hiding places, general attraction of young and adult members of the same species by a familiar odour, marking of

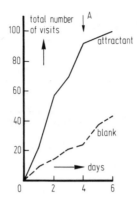

Figure 6. Results of one of the tests in the basin in Rijswijk (see Figure 5) obtained during screening of a variety of muskrat lures. Attractant No. 29 ("muskoil") was obtained (December 5–11, 1980) by extraction of preputial glands with ethanol.

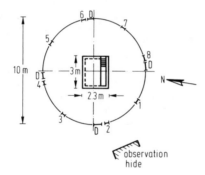

Figure 7. Schematic representation of the "olfactometer" consisting of a round muskrat pen of iron gauze (diameter 10 m; height 2.20 m). An iron swimming pool was placed in the middle. Key: 1–8, counting cages with encoder box (see Figure 8); and D, door.

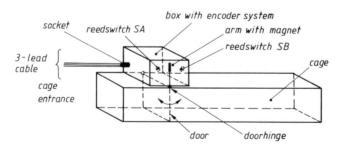

Figure 8. Cage with encoder box in which time, number and duration of visits of muskrats are signaled and transmitted to a recorder.

territory as well as stimulation ("priming") of endocrine processes, etc..

Among the few clues found in the literature which might indicate a function, are those of Akkerman (22) who observed that a
male muskrat was "scent-marking" a female during courtship and
mating which occurs while they are swimming. He also found that
faecal piles are highly important in muskrat communication, as
Müller-Schwarze recently described for beavers (23). In spring
the faecal piles of muskrats are found where they leave the water.
These piles are produced where population densities are high,
which again parallels the situation in beavers (23).

The muskrat secretes its musk during scent marking through
the penis. The soles of the hind feet pick up the fluid and throw
it behind the animal (22).

Chemical composition and seasonal changes of musk gland secretion

The preputial (musk) gland of the muskrat contains a number of macrocyclic ketones, such as isomers of civetone (cis-9-
cycloheptadecenone), the main odourous component of the secretion
of the civet. In the muskrat the main macrocyclic component is
the cis-5-isomer of civetone (Figure 9). The glands of both animals also contain a number of other macrocyclic compounds and a
variety of fatty acids (24,25,26) which could be precursors of
macrocyclic ketones and alcohols as proposed by Stevens in 1945
(25).

The vast literature on the muskrat mentioned before, as summarized by Errington (27) and Hoffmann (19) mainly concerns
ecology, taxonomy, anatomy, parasitology and economy. Only a few
chemical studies have been carried out and the interest in the
biological functions of the components found in the gland secretion seemed to be negligible.

In 1973, Van Dorp et al. found a number of interesting macrocyclic compounds, including some with both double and triple
unsaturated bonds (26), but made no attempt to determine their
biological significance. They compared the macrocyclic ketones and
the fatty acids found in civet and muskrat gland and concluded
that there is evidently no correlation between the macrocyclic
ketone composition and the fatty acid composition. For example,
cycloheptadecanone was the most prominent peak in the muskrat
gland (41% of the macrocyclic ketones), whereas the stearic acid
from which it could have been formed, was among the minor components (3%) of the fatty acids.

We analysed ethanolic extracts of preputial glands of muskrats
collected during a full annual cycle by one trapper, in Zeeuwsch-
Vlaanderen, near the Belgian border. The majority of the glands
were taken from adult males, but some glands of adult females and
young females and males were also collected and analysed separately (GC and MS; SE-30 capillary column).

Tables I and II compare the compounds which we have found so
far with those found by Van Dorp et al. Although we have not yet

been able to definitely establish the position of most of the
double or triple bonds, the data show that probably nine ketones
were found by both groups, whereas two components identified by
Van Dorp et al. have not yet been found by us and four of our
compounds were not found by the other group.

Again with some reserve regarding the position of the double
bonds, Table II shows that we found all the acids mentioned by Van
Dorp et al., but, in addition, nine more acids lacking from their
list.   .

There is, however, a great similarity with regard to the
question which were the major and which the minor components in
the two groups of compounds.

We cannot, however, support the opinion that there is no
correlation between the ketones and fatty acids. Analyses made
over a one year period seem to indicate that the percentage of
macrocyclic ketones is often high when that of the acids is low
and vice versa. This would support Stevens' hypothesis, which was
challenged by Van Dorp et al. (26), that the ketones are formed
by cyclization and decarboxylation of fatty acids (25).

As an example, Figure 10 shows the seasonal changes in the
relative percentage of all the 17-membered ring ketones together
and those of the C_{18} acids and esters. These data not only support the biogenetic relationship between acids and macrocyclic
ketones but also suggest that the compounds with the musk odour
are mainly produced during fall and winter. The percentage of the
musk compounds drops sharply during the first three months of the
year and increases again in spring.

The analyses of the female glands showed that these also
contain some macrocyclic compounds but that their relative percentages are much lower than those found in the males and that,
correspondingly, the percentage of fatty acids is larger.

We wish to thank Unilever, Vlaardingen for supplying samples of
macrocyclic ketones and Dr. P.E.J. Verwiel and Ing. A. Lakwijk for
carrying out GC/MS experiments. The work was supported by a grant
from the Dutch Departments of Wildlife Management and Public Works.
We also thank Dr. W.J. Doude van Troostwijk and his collaborators
of the Committee for Muskrat Control for their valuable help and
advices.

Structure Elucidation of Periplanone-A

The names periplanone-A and periplanone-B were first coined by
Ritter and Persoons in 1974 at the International Congress of
Pesticide Chemistry in Helsinki, in a general paper on pheromone
research in the Netherlands. In this paper the isolation of a
number of pheromones were described, including the two peripla-
nones, both sex pheromones, present in the excreta of the female
American cockroach, Periplaneta americana. The elementary formulae
of these two sesquiterpenoids, $C_{15}H_{20}O_2$ and $C_{15}H_{20}O_3$, were given
and the carbon skeleton of periplanone-B, a germacrane derivative,
was revealed. Unfortunately, we were informed, one year after the

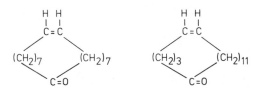

Figure 9.   Structures of the main macrocyclic ketones of the civet (left) and the muskrat (right).

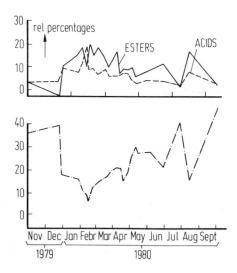

Figure 10.   Variation with season of the proportion of macrocyclic C$_{17}$ ketones and of their possible precursors (bottom), the C$_{18}$ acids and their esters (top). Sum of all compounds detected by the GC analysis is 100%.

Table I

MACROCYCLIC KETONES IN THE PREPUTIAL GLAND OF THE MUSKRAT

Compound	V. Dorp et al.	Ritter et al.
Cyclotridecanone	−	+
Cyclopentadecanone	+++	+++
Cyclopentadecenone, (Z)-5-	+++	+++
Cyclopentadecynone, 5-*	+	+
Cyclohexadecanone	+	+
Cyclohexadecenone*	−	+
Cycloheptadecanone	+++	+++
Cycloheptadecenone, (Z)-5-	+++	+++
Cycloheptadecenone, (Z)-7-*	++	++
Cycloheptadecadienone, (Z,Z)-5,11-*	++	−
Cycloheptadecynone, 5-*	+	−
Cycloheptadecen-5-ynone, (Z)-7-*	++	++
Cyclononadecanone	+	+
Cyclononadecenone, (Z)-9-*	−	+
Cyclononadecenone*	−	+

* Position of double bond(s) not yet established in our work.

+++ ≥ 10% relative of the macrocyclic ketones
++   > 1%   "   "   "   "   "
+   ≤ 1%   "   "   "   "   "
−   not found, or not identified

Table II

FATTY ACIDS IN THE PREPUTIAL GLAND OF THE MUSKRAT

Fatty acid		V. Dorp et al.	Ritter et al.
Number of C-atoms	Number of double bonds*		
12	0	−	+
12	1	−	+
14	0	++	++
14	1	−	+
16	0	+++	+++
16	1	++	++
16	1	−	++
17	0	−	+
18	0	++	++
18	1	+++	+++
18	1	−	+
18	2	++	++
19	1	−	+
20	0	−	+
20	1	+	+
20	2	−	+

* Position of double bond(s) not yet established in our work.

+++ ⩾ 10% relative of the fatty acids
++ > 1%   "   "   "   "   "
+   ⩽ 1%   "   "   "   "   "
−   not found, or not identified

congress, that none of the symposium papers could be published
and that we had to seek publication elsewhere. This we did in
the Netherlands Journal of Zoology in 1975 (28). Meanwhile we had
published the experimental details in our communication to the
Royal Dutch Academy of Sciences (29). A similar paper had been
rejected for publication in 1973, first by Science, and then by
Experientia.

We suspect that the reluctancy of authoritative journals
outside Holland to publish our data resulted from fear of repeat-
ing the unfortunate events after the premature publication of a
wrong structure for this sex pheromone by eminent American
scientists in 1963 (30), who had to withdraw their structure two
years later (31).

A few years ago the complete structure of periplanone-B,
apart from its stereochemistry, was elucidated and published by
our research team (32,33,34) with Persoons as the principal
investigator, who included this work in his doctoral thesis (35),
and shared the Royal Dutch Shell Prize for 1978 with Ritter for
this work and the structure elucidation of faranal, the trail
pheromone of the Pharaoh's ant, Monomorium pharaonis (36,37).

The determination of the absolute configuration of periplanone-B by American scientists in cooperation with Persoons (38)
and the confirmation of the structure by the elegant synthesis
of Still at the Columbia University (39) followed in 1979. The
planar and stereostructure of periplanone-B are shown in
Figure 11.

We can now report the structure elucidation of the other
isolated compound: periplanone-A.

It is, as we mentioned in earlier papers (28,29) an unstable
sesquiterpenoid with the general formula $C_{15}H_{20}O_2$. Its structure,
like that of periplanone-B, is rather complicated compared with
those of other known sex pheromones. It could be deduced from
comparison of its NMR, UV, IR and mass spectra with those of an
inactive rearrangement product (Figure 12), the carbon skeleton
of which resembles that of periplanone-B, apart from being bicyclic instead of having a 10-membered ring. This rearrangement
product, with the same elementary formula as periplanone-A is
readily formed from it when kept in NMR capillaries in carbon
disulfide or hexane. On the basis of NMR coupling constants and
its other spectroscopical data it would have the configuration
shown in Figure 13. It can be separated from periplanone-A by GC,
using DEGS or OV 101 as the stationary phase.

Scrutinous analysis of the NMR spectra of mixtures containing decreasing amounts of periplanone-A with the NMR spectrum
of the rearrangement product yielded the NMR signals attributable
to periplanone-A. Assuming that signals with similar chemical
shifts and patterns in the spectra of periplanone-A and its
rearrangement product arose from identical partial structures and
subsequent combination of these partial structures, together with
a consideration of chemical shifts and coupling constants, lead

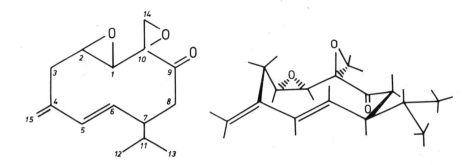

Figure 11. Structure of periplanone-B: (1Z, 5E)-1,10(14)-diepoxy-4(15), 5-germacradiene-9-one planar structure (left) and stereostructure of the natural enantiomer (1R, 2R, 5E, 7S, 10R) (right).

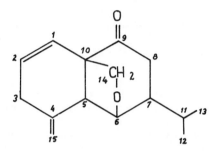

Figure 12. Structure of the rearrangement product of periplanone-A.

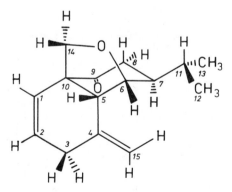

Figure 13. Expected configuration for the rearrangement product of periplanone-A.

to the structure for periplanone-A, shown in Figure 14. According
to IUPAC nomenclature it is 7-methylene-4-isopropyl-12-oxa-
tricyclo $[4.4.2.0^{1,5}]$-9-dodecene-2-one.

The structure of periplanone-A as well as that of its rearrangement product still requires confirmation by synthesis, but
all their spectral data are in complete agreement with the proposed structures. A paper in which all these details are given is
in press (40).

Periplanone-A and periplanone-B have been found to occur in
a ratio of 1:10 in faecal material of Periplaneta americana, but
only periplanone-B has been found in intestinal tracts. This could
suggest that periplanone-B is a genuine sex pheromone, whereas
periplanone-A could be a biologically active transformation
product, which in turn can isomerize into a more stable but inactive compound. We have given hypothetical schemes for such a
transformation elsewhere (32,40).

This would involve a reduction of the 1(2)-epoxigroup of
periplanone-B, followed by a transannular reaction of the other
epoxigroup (the exo-epoxide at C_{10}) with the trans-double bond of
a hypothetical intermediate at C_5, which would yield periplanone-A.

Via another hypothetical intermediate, the 5-cis isomer,
periplanone-A could then be transformed into a stable, but inactive compound (Figure 15).

Recently electrophysiological experiments by Sasz have
yielded results which would indicate that not only periplanone-B,
but also periplanone-A could be a genuine pheromone. With single
cell recording techniques he demonstrated the presence of many
olfactory hairs on the antennae of Periplaneta americana that were
very sensitive to periplanone-A, whereas others were sensitive for
periplanone-B (41).

Although periplanone-A is a very potent sex pheromone, its
practical application might meet more difficulties than that of
periplanone-B and would require a solution of the problems related
to its instability.

We wish to acknowledge the contribution of W.J. Nooijen to
the isolation and of P.E.J. Verwiel to the NMR analyses of periplanone-A.

The Sex Pheromone of the Beet Armyworm, Spodoptera exigua

The beet armyworm, Spodoptera exigua, is a rather recent pest in
Holland, where it is now called the Florida moth. After it was
accidentally introduced into the Netherlands in 1976, it immediately became a grave pest in greenhouses, mainly for chrysanthemums
and gerberae, but to a lesser extent also for egg-plants. As
application of insecticides such as synthetic pyrethroids, initially effective, soon proved to have become inadequate, other
methods were needed, and pheromones were investigated as possible
alternatives.

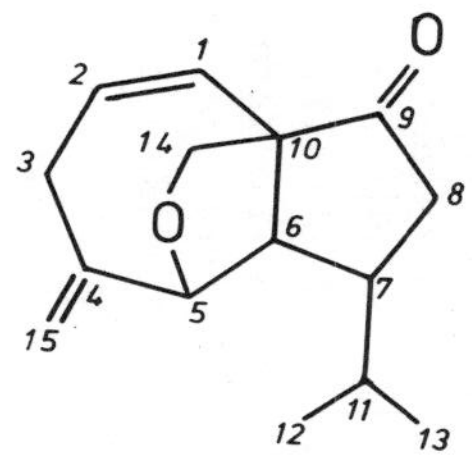

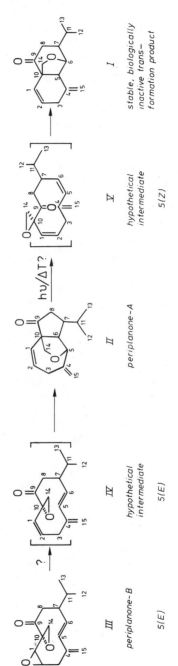

Figure 14. Proposed structure of periplanone-A.

III	IV	II	V	I
periplanone-B	hypothetical intermediate	periplanone-A	hypothetical intermediate	stable, biologically inactive trans-formation product
5(E)	5(E)		5(Z)	

Figure 15. Hypothetical scheme for the transformation of periplanone-B into periplanone-A and of periplanone-A into the stable, biologically inactive product.

Brady and Ganyard identified (Z,E)-9,12-tetradecadienyl
acetate in this moth (42), but trapping experiments with this compound, alone or in combination with (Z)-9-tetradecenyl acetate
were unsuccessful (43,44). A joint effort to reinvestigate this
problem was therefore started by research teams in Delft and
Wageningen.

Fractions obtained from abdominal tips, excised glands or
trapped air volatiles obtained from virgin females were tested by
electroantennography and a behavioural bioassay (Van der Kraan,
publication in preparation). The latter used wind tunnels with
funnels upwind and a common chamber downwind. Test samples were
placed in the funnels, and the behaviour of male moths, released
in the middle of the tunnels, as well as the numbers trapped in
the funnels, were examined.

Of 150 model compounds screened by the EAG technique, only
the (Z,E)-9,12-14:Ac isolated by Brady and Ganyard gave a clear
positive response.

The behavioural assay of fractions eluted from silicagel
with, respectively, hexane, 10% ether in hexane, 50% ether in
hexane and ether, indicated that the activity was confined to the
fraction containing C_{14} acetates (the 10% ether fraction). Combination with other fractions did not increase the activity.

Analyses by mass spectrometry, ozonolysis and retention
indices revealed the presence of the (Z,E)-9,12 and the (Z)-9 compounds mentioned earlier, but, in addition, also of the (Z,Z)-
9,12 di-unsaturated, the (Z)-11 monounsaturated and the saturated
C_{14} acetates. The details of the analyses and the different ratios
found in abdominal tips, excised glands and air volatiles are
given in a recent communication (45).

The behavioural assay indicated that the (Z,E)-9,12 and the
(Z)-9 components appeared to be necessary for upwind search
behaviour, while the (Z,Z)-9,12 compound is important for short
range courtship behaviour, like hair pencil display and copulation
attempts.

Field experiments were carried out by T.C. Baker near Riverside in California, who used Pherocon-2 sticky traps and polyethylene caps as dispensers.
The preliminary results show that traps with mixtures of the four
unsaturated acetates collected as an average slightly more males
than traps with live females did. Remarkably, mixtures of the two
compounds found to be ineffective by Mitchell and Doolittle (44)
caught only slightly fewer males than these two live virgin
females did.

While our most recent communication was in press (45), a
publication on the same subject by Tumlinson, Mitchell and Sonnet
appeared (46). They isolated 11 compounds from female Spodoptera
exigua, but their field tests demonstrated that a mixture of only
two of them, (Z,E)-9,12-tetradecadienyl acetate and the free
alcohol, (Z)-9-tetradecenol, in a ratio of 5:4, were needed for
attractancy. Such a mixture was equal in activity to live females,

to the total hexane wash of females and to a mixture of all
11 compounds in attracting wild males in the field.

   To obtain attraction comparable to that of living females,
the authors used about 45 mg of their mixture in polyethylene
vials, which in our opinion, is an excessively high dose. We
always used doses below the 1 mg level.

   Table III compares the results obtained by Mitchell and
Doolittle in 1976 (44), by Tumlinson et al. in 1981 (45) and by
Persoons et al. (46). It lists the 11 compounds found by
Tumlinson et al.. Of the five acetates which we investigated,
four were also found by the other group. The results of the field
tests differ widely. Whereas Mitchell and Doolittle found no
attractancy in the two-component mixture, Bakers' field tests
found a similar mixture to be almost as attractive as live
females.

   Very recently, Baker compared the two-component mixture of
Tumlinson et al. with our four-component mixture in field tests
near Riverside. Although, due to low population densities, the
catches were rather poor, the results are clearly in favour of
the four-component mixture (Table IV).

   We consider all these field tests still to be preliminary.
They will have to be repeated on a larger scale and in different
locations and countries before a complete picture can be gained.
The work on Spodoptera exigua in Holland was done in collaboration
with W.J. Nooijen of the Delft group and C. van der Kraan and
S. Voerman of the Institute for Pesticide Research in Wageningen.
We also gratefully acknowledge the field tests by T.C. Baker in
Riverside and the suggestions and critical comments by A.K. Minks
of the Institute for Plant Protection in Wageningen.

Table III

RESPONSE OF WILD SPODOPTERA EXIGUA MALES TO SYNTHETIC PHEROMONE
IN POLYETHYLENE CAPS

	Mitchell and Doolittle 1976			Tumlinson et al. 1981			Persoons et al. 1981		
	mg	mg	3♀	µl	µl	4♀	µg	µg	2♀
14:Ac				+			+		
(Z)-7-14:Ac				+					
(E)-9-14:Ac				+			(+)		
(Z)-9-14:Ac	+	3	5	+	12.5		+	450	300
(Z)-11-14:Ac							+	45	
(Z,E)-9,12-14:Ac	+	7	5	+	25	25	+	450	600
(Z,Z)-9,12-14:Ac				+			+22.5		
14:OH				+					
(E)-9-14:OH				+					
(Z)-9-14:OH				+	20	20			
(Z,E)-9,12-14:OH				+	10				
(Z,Z)-9,12-14:OH				+					
x̄ no. ♂/trap/night	0.0	0.0	5.3	19.7	26.7	12.0	8.5	7.8	8.0

+ compound detected

Table IV

RESPONSE OF A LOW POPULATION OF WILD SPODOPTERA EXIGUA MALES TO
SYNTHETIC PHEROMONE IN POLYETHYLENE CAPS.
COMPARATIVE FIELD TEST OF "TUMLINSON" AND "PERSOONS" MIXTURES,
TESTED BY BAKER

	Tumlinson		Persoons		
	µg	µg	µg	µg	µg
(Z)-9-14:Ac			600	200	1000
(Z)-11-14:Ac			60	20	100
(Z,E)-9,12-14:Ac	500	5000	400	200	750
(Z,Z)-9,12-14:Ac			20	10	37.5
(Z)-9-14:OH	400	4000			
x̄ no. ♂/trap/night	0.4	0.0	1.8	2.0	1.5

Literature cited

1. Mykytowycz, R., in: Ritter, F.J., Ed.; "Chemical Ecology: Odour Communication in Animals"; Elsevier: Amsterdam, 1979; pp. 105-115.
2. Beauchamp, G.K.; Doty, R.L.; Moulton, D.G.; Mugford, R.A. J. Chem. Ecol. 1979, 5, 301-305.
3. Katz, R.A.; Shorey, H.H. J. Chem. Ecol. 1979, 5, 299-301.
4. Rutowski, R.L. J. Chem. Ecol. 1981, 7, 481-484.
5. Claus, R. "Pheromone bei Säugetieren unter besonderer Berücksichtigung des Ebergeruchstoffes und seiner Beziehung zu anderen Hodensteroiden", Paul Parey: Hamburg, 1979.
6. Müller-Schwarze, D.; Mozell, M.M., Eds.; "Chemical Signals in Vertebrates"; Plenum: New York, 1977.
7. Müller-Schwarze, D.; Silverstein, R.M., Eds.; "Chemical Signals. Vertebrates and Aquatic Invertebrates"; Plenum: New York, 1980.
8. Ritter, F.J., Ed.; "Chemical Ecology: Odour Communication in Animals"; Elsevier: Amsterdam, 1979.
9. Shorey, H.H. "Animal Communication by Pheromones"; Academic Press: New York, 1978.
10. Birch, M.C., Ed.; "Pheromones"; Elsevier: New York, 1974.
11. Christiansen, E.; Døving, K., in: Hanson, L.; Nilson, B., Eds.; "Biocontrol of Rodents"; Swed. Natl. Res. Council: Stockholm, 1975; 221-227.
12. Marsh, R.E.; Howard, W.E. Pest Control 1979, 7, 22-23.
13. Stacewycz-Sapuntzakis, M.; Gawienowski, A.M. J. Chem. Ecol. 1977, 3, 411-417.
14. Christian, J.J. Am. J. Physiol. 1955, 182, 292-300.
15. Davis, D.E.; Christian, J.J. Nav. Med. Res. Inst. Res. Rep. 1957, 15, p 311.
16. Bronson, F.H.; Caroom, D. J. Reprod. Fertil. 1971, 25, 279-282.
17. McKinney, T.D.; Christian, J.J. Proc. Soc. exp. Biol. Med. 1970, 134, 291-293.
18. Müller-Schwarze, D., in: Eisenberg, J.F.; Kleiman, D.G., Eds.; "The Behavior of Mammals"; Amer. Soc. Mammalogists Publ. (in press).
19. Hoffmann, M. Z. angew. Zool. 1976, 63, 187-239.
20. Williams, R.M. J. Wildlife Management 1951, 15, 117-118.
21. Brüggemann, I.E.M.; Ritter, F.J.; Gut, J. "Onderzoek naar signaalstoffen voor de muskusrat"; 1981; TNO report CL 81/24.
22. Akkermann, R. Z. angew. Zool. 1975, 62, 281-338.
23. Müller-Schwarze, D.; Heckman, S. J. Chem. Ecol. 1980, 6, 81-95.
24. Stevens, P.G.; Erickson, J.L.E. J. Am. Chem. Soc. 1942, 64, 144-147.
25. Stevens, P.G. J. Am. Chem. Soc. 1945, 67, 907-909.
26. Van Dorp, D.A.; Klok, R.; Nugteren, D.H. Rec. Trav. Chim. Pays-Bas 1973, 915-928.

27. Errington, P.L. "Muskrat Populations"; Iowa State Univ. Press: Ames, 1963.
28. Ritter, F.J.; Persoons, C.J. Neth. J. Zool. 1975, 261-275.
29. Persoons, C.J.; Ritter, F.J.; Lichtendonk, W.J. Proc. Kon. Ned. Akad. Wetensch. Series C 1974, 77, 201-204.
30. Jacobson, M.; Beroza, M.; Yamamoto, R.T. Science 1963, 139, 48-49.
31. Jacobson, M.; Beroza, M. Science 1965, 147, p 749.
32. Persoons, C.J.; Ritter, F.J., in: Ritter, F.J., Ed.; "Chemical Ecology: Odour Communication in Animals"; Elsevier: Amsterdam, 1979, pp. 225-236.
33. Persoons, C.J.; Verwiel, P.E.J.; Talman, E.; Ritter, F.J. J. Chem. Ecol. 1979, 5, 221-236.
34. Talman, E.; Verwiel, P.E.J.; Ritter, F.J.; Persoons, C.J. Israel J. Chem. 1978, 17, 227-235.
35. Persoons, C.J. "Structure elucidation of some insect pheromones; a contribution to the development of selective pest control agents"; Thesis, Wageningen, 1977.
36. Ritter, F.J.; Brüggemann, I.E.M.; Verwiel, P.E.J.; Talman, E.; Stein, F.J.; Persoons, C.J., in: Kloza, M., Ed. in chief; "Regulation of Insect Development and Behaviour"; Wroclaw Techn. Univ. Press: Wroclaw 1981; pp. 871-882.
37. Ritter, F.J.; Brüggemann-Rotgans, I.E.M.; Verwiel, P.E.J.; Persoons, C.J.; Talman, E. Tetrahedron Letters 1977 (30), 2617-2618.
38. Adams, M.A.; Nakanishi, K.; Still, W.C.; Arnold, E.V.; Clardy, J.; Persoons, C.J. J. Am. Chem. Soc. 1979, 101, 2495-2498.
39. Still, W.C. J. Am. Chem. Soc. 1979, 101, 2493-2495.
40. Persoons, C.J.; Verwiel, P.E.J.; Ritter, F.J.; Nooijen, W.J. J. Chem. Ecol. (in press).
41. Sasz, H., in: Van der Starre, H., Ed.; "Olfaction and Taste VIII"; Int. Retrieval Ltd.: London, 1980; p 194.
42. Brady, U.W.; Ganyard, M.C. Ann. Ent. Soc. Am. 1972, 65, 898-899.
43. Campion, D.G. Meded. Fac. Landbouw Gent 1975, 40, 283-292.
44. Mitchell, E.R.; Doolittle, R.E. J. Econ. Entomol. 1976, 69, 324-326.
45. Persoons, C.J.; Van der Kraan, C.; Nooijen, W.J.; Ritter, F.J.; Voerman, S.; Baker, T.C. Entomol. exp. et appl. 1981, 30, 98-99.
46. Tumlinson, J.H.; Mitchell, E.R.; Sonnet, P.E. J. Environ. Sci Health 1981, A16, 189-200.

RECEIVED February 24, 1982.

Development of Microencapsulated Pheromone Formulations

D. R. HALL and B. F. NESBITT—Overseas Development Administration, Tropical Products Institute, London WC1X 8LU, England

G. J. MARRS and A. St. J. GREEN—ICI Ltd., Plant Protection Division, Bracknell, Berkshire RG12 6EY, England

D. G. CAMPION and B. R. CRITCHLEY—Overseas Development Administration, Centre for Overseas Pest Research, London W8 5SJ, England

Microencapsulation provides a means of formulating synthetic pheromones for application to large areas in the control of insect populations by disruption of pheromone-mediated communication. Microencapsulated formulations are easily manufactured by known methods of interfacial polymerisation, they can be applied to large areas with conventional spray equipment and they possess numerous variables that can be used to control the release characteristics. The physicochemical properties of these formulations have been studied by determination of the residual amount of pheromone remaining after varying lengths of exposure: incorporation of light-stable analogues into the formulations and comparison of the results from samples exposed to and shielded from direct sunlight showed that both pheromone and formulation suffered from degradation in sunlight. Various types of antioxidant were tested, but only phenylene diamines were effective in reducing the degradation. N,N'-Disecondary phenylene diamines could not be encapsulated, but the corresponding N,N'-ditertiary derivatives in combination with a dye + carbon black UV screener gave formulations with greatly improved stability in sunlight. Such formulations of the pheromones of *Pectinophora gossypiella* and *Spodoptera littoralis* caused high levels of communication disruption of these species in the field.

Although pheromone traps are now widely used for monitoring insect populations, the use of synthetic pheromones in control of insect pests by disruption of pheromone-mediated communication has been much slower to develop, at least in part because of the difficulties of formulation. There are three main types of formulation currently available - hollow fibres, plastic laminates and microcapsules. For several years, the UK Overseas

Development Administration and ICI Ltd. have been involved in a
collaborative project to develop microencapsulated formulations
of lepidopterous sex pheromones for control of insect pests by
mating disruption.

Microencapsulated formulations prepared by interfacial
polymerisation were favoured for three main reasons:
  i) they are easily manufactured on a large scale using
     known technology (1);
 ii) they are easily applied over large areas with
     conventional spray equipment;
iii) they possess numerous variables that can be man-
     ipulated to control the release characteristics
     (2), e.g. capsule wall thickness, capsule wall
     composition, capsule size and internal composition.

Persistence Studies

     In most of the work, the physicochemical behaviour of the
formulations was studied by gas chromatographic estimation at
intervals of the residual pheromone remaining in formulations
sprayed onto filter papers. The latter were of the silicone-
treated "phase-separating" type as these best simulated a leaf
surface. When the formulations were exposed in a laboratory
wind-tunnel there was little pheromone loss other than by
release, at least for the monounsaturated acetates used in most
of the preliminary work, and such analyses provided accurate
information on release rates under these conditions. However,
use of this technique in the field showed that loss of pheromone
was very much more rapid than under comparable conditions of
temperature and windspeed in the wind-tunnel (2). These results
were taken to indicate that there was significant loss of
pheromone by degradation under field conditions.

     The extent of pheromone degradation under field conditions
was investigated with a microencapsulated formulation containing
a saturated hydrocarbon and acetate (octadecane and tetradecyl
acetate (14:Ac)), the corresponding monounsaturated hydrocarbon
and acetate ((Z)-4-octadecene and (Z)-9-tetradecenyl acetate
(Z9-14:Ac)) and a diunsaturated acetate ((Z,E)-9,11-tetradeca-
dienyl acetate (ZE9,11-14:Ac)), chosen so that all the components
had similar volatilities. On exposure to sunlight, loss of the
diene was more rapid than loss of the monounsaturated components
which in turn disappeared faster than the saturated components
(Fig. 1). All components disappeared at a similar, slower rate
when shielded from direct sunlight.

     Thus subsequent experiments were carried out with two
important modifications.
  i) Formulations were sprayed onto filter papers, half
     of which were exposed to direct sunlight while the
     other half were exposed to the same conditions of
     temperature, windspeed and humidity, but shielded

from direct sunlight; rates of disappearance of
encapsulated components in exposed and shielded
samples were then compared.
ii) Light-stable analogues were incorporated into the
formulations, e.g. the corresponding saturated
acetates in formulations of unsaturated acetates.

Figure 2 shows the four curves obtained from such an
experiment carried out in Egypt with a polyurea microencapsulated
formulation of Z9-14:Ac and 14:Ac. The extent of degradation of
the unsaturated Z9-14:Ac is indicated by the difference between
the curve for this component and the corresponding curve for the
light-stable 14:Ac. In this experiment - cf. Fig. 1 from an
experiment carried out in London during winter - there is even
significant degradation of the unsaturated component in the
formulation shielded from direct sunlight although still exposed
to reflected light. The difference between the curve for the
disappearance of the 14:Ac in the exposed samples and that for
the shielded samples is a measure of the degradation of the
microcapsules themselves in direct sunlight.

Stabiliser Studies

Rapid degradation of simple mono-olefins in sunlight was
unexpected as they show absorbance maxima well below the 300 nm
cut-off observed for natural sunlight. However, the effect was
easily demonstrated by exposing mixtures of unsaturated compounds
and light-stable analogues as thin films in open petri dishes and
analysing the relative amounts of the two components remaining.
Various antioxidants were tested in this system, including
hindered phenols, gallate esters and phenylene diamines. Only
the latter had any significant effect in reducing degradation,
and the best were those with additional aromatic substituents on
nitrogen. Table I shows the results of an experiment in which
mixtures of 1 mg each of Z9-14:Ac and 14:Ac with 0.2 mg of the
antioxidant were exposed as thin films in open 4 cm-diameter
petri dishes on the laboratory roof in London. After exposure
for 10 daylight hours the amounts of Z9-14:Ac and 14:Ac remaining
were determined by gas chromatography. The ratio between the
amount of the unsaturated component and that of the saturated
component remaining provides a measure of the extent of the
degradation of the former since the rates of evaporation of the
two components will be similar. A final ratio of 1.0 (the
starting ratio) indicates no degradation, and a ratio of 0.0
indicates complete degradation of the unsaturated component
during the experiment.
Of the commercially-available phenylene diamines tested, the
N,N'-diphenyl compound is an insoluble, high melting-point,
crystalline solid, whereas the unsymmetrically substituted
N-phenyl-N'-(2-octyl) derivative is a liquid, readily miscible

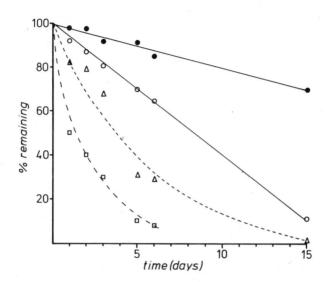

Figure 1. Exposure of unstabilized microcapsule formulation on TPI roof in London. Key: octadecane, 14:Ac, (○); (Z)-4-octadecene, Z9-14:Ac, (△); ZE9,11-14:Ac, (□); and all components shielded, (●).

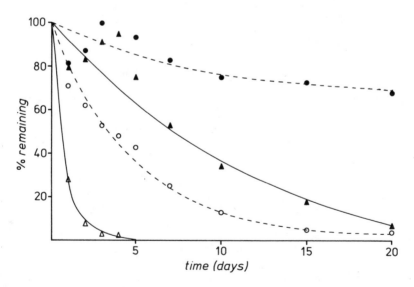

Figure 2. Exposure of unstabilized microcapsule formulation in Egypt. Key: Z9-14:Ac, exposed (△) and shielded (▲); and 14:Ac, exposed (○) and shielded (●).

TABLE I

STABILISATION OF Z9-14:Ac BY VARIOUS PHENYLENE DIAMINE

ANTIOXIDANTS

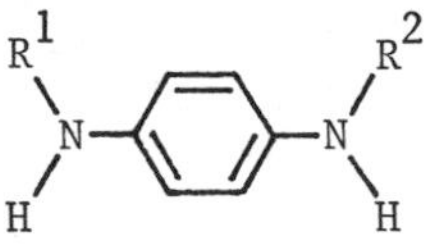

antioxidant		$\dfrac{Z9-14:Ac}{14:Ac}$
R^1	R^2	
$-C_8H_{17}$	$-C_8H_{17}$	0.62
$-Ph$	$-C_8H_{17}$	0.90
$-Ph$	$-Ph$	0.92
no antioxidant		0.47

1 mg Z9-14:Ac + 1 mg 14:Ac + 0.2 mg
antioxidant exposed in open petri dishes
on TPI roof for 10 daylight hours

with typical non-polar pheromones. This compound is commercially available as the antiozonant UOP 688, but, as a secondary amine, it cannot be encapsulated using interfacial polymerisation since it reacts with one of the monomers. Fully substituting the nitrogens in the UOP 688 caused some reduction in the stabilising effect on pheromone degradation (Table II), but the resulting tertiary amines could be encapsulated and still retained useful stabilising activity (Table III).

A range of substituents on nitrogen was tested (Table IV) and all the derivatives possessed stabilising activity. Further work was carried out with the N,N'-dimethyl compound as this was most conveniently prepared (1).

Various UV screeners were also tested for their stabilising activity, including hydroxybenzophenones, aminobenzoates, dyes and carbon black. Only combinations of dyes and carbon black showed any effect, the best being the commercially-available Waxoline Black (Table V).

TABLE II

STABILISATION OF Z9-14:Ac BY VARIOUS PHENYLENE DIAMINE
ANTIOXIDANTS

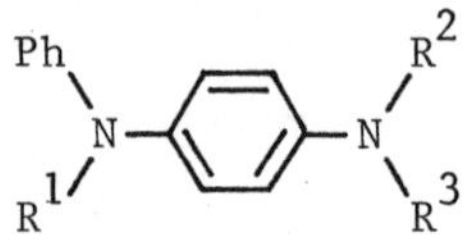

antioxidant				$\dfrac{Z9\text{-}14\text{:Ac}}{14\text{:Ac}}$
R^1	R^2	R^3	amount	
$-CH_3$	$-CH_3$	$-C_8H_{17}$	5 mg	0.93
$-H$	$-H$	$-C_8H_{17}$	1 mg	0.75
$-H$	$-H$	$-Ph$	1 mg	0.72
no antioxidant				0.28

1 mg Z9-14:Ac + 1 mg 14:Ac + antioxidant exposed in
open petri dishes on TPI roof for 20 daylight hours

TABLE III

STABILISATION OF Z9-14:Ac BY N,N'-DIMETHYL UOP 688

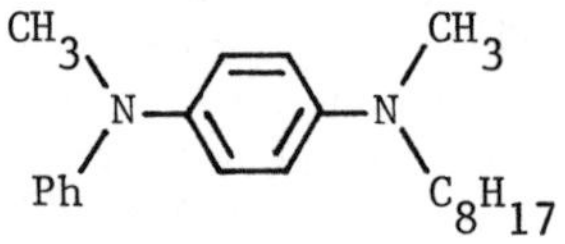

exposure time	$\dfrac{Z9\text{-}14\text{:Ac}}{14\text{:Ac}}$
25 hr	0.92
65 hr	0.78
113 hr	0.60
no antioxidant (25 hr)	0.14

1 mg Z9-14:Ac + 1 mg 14:Ac + 5 mg antioxidant
exposed in open petri dishes on TPI roof

TABLE IV

STABILISATION OF Z9-14:Ac BY DERIVATIVES OF UOP 688

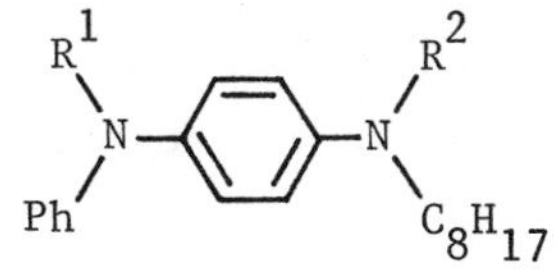

$$R^1 \diagdown \underset{Ph}{\overset{}{N}} - \bigcirc - \underset{C_8H_{17}}{\overset{R^2}{N}}$$

antioxidant		Z9-14:Ac
R^1	R^2	14:Ac
-H	-H	0.92
$-CH_3$	$-CH_3$	0.91
$-C_2H_5$	$-C_2H_5$	0.89
$-C_{10}H_{21}$	$-C_{10}H_{21}$	0.83
$-CH_2Ph$	$-CH_2Ph$	0.86
$-CH_2CH=CH_2$	$-CH_2CH=CH_2$	0.89
$-CH(CH_3)_2$	$-CH_3$	0.87
-H	$-COCH_3$	0.60
no antioxidant		0.07

1 mg Z9-14:Ac + 1 mg 14:Ac + 1 mg antioxidant exposed in open petri dishes on TPI roof for 60 daylight hours

TABLE V

STABILISATION OF Z9-14:Ac

stabiliser	Z9-14:Ac 14:Ac
Waxoline Black	0.82
UOP 688	0.80
no stabiliser	0.38

1 mg Z9-14:Ac + 1 mg 14:Ac + 1 mg stabiliser exposed in open petri dishes on TPI roof for 20 daylight hours

Exposure Trials

The above studies were carried out using the simple, exposed thin-film system. Microencapsulated formulations of Z9-14:Ac and 14:Ac were made up containing the N,N'-dimethyl UOP 688 and/or Waxoline Black, and tested in exposure experiments in Egypt. The N,N'-dimethyl UOP 688 alone stabilised the pheromone but seemed to have little effect on the degradation of the capsule walls (Fig. 3), while the Waxoline Black alone had a greater effect on the stability of the capsules than on that of the pheromone (Fig. 4). Combination of the two had a synergistic effect on the stabilities of the microcapsule wall and the pheromone (Fig. 5).

Field Trials

Small-scale field trials with fully stabilised micro-encapsulated formulations were carried out in Egypt during 1980 on two cotton pests - pink bollworm, *Pectinophora gossypiella*, and the Egyptian cotton leafworm, *Spodoptera littoralis*. Plots of 10 m x 10 m were sprayed with the formulation and there were three replicates of each treatment. A plastic funnel trap (3) baited with synthetic pheromone was positioned at the centre of each plot. The effectiveness and persistence of communication disruption were measured by the reduction in catches of male moths in the traps in the treated plots relative to the catches in similar traps in untreated, control plots.

Reduction in catches of *P. gossypiella* lasted for about two weeks when the formulation of the pheromone (a 1:1 mixture of (Z,E)- and (Z,Z)-7,11-hexadecadienyl acetate) was applied at a level of 5 gm a.i. per ha (Fig. 6). When the formulation was applied at 20 gm a.i. per ha, 97% or greater reduction in trap catches was maintained for at least four weeks even when moth catches in the untreated plots were very high (Fig. 7). Similar results were obtained with a hollow fibre formulation (4) of the pheromone applied at similar rates by hand (Figs. 8 and 9).

The major component of the pheromone of *S. littoralis* is the conjugated diene (Z,E)-9,11-tetradecadienyl acetate which is very susceptible to degradation by heat and sunlight (5). The micro-encapsulated formulation of this diene had to be applied at the higher level of 40 gm a.i. per ha to achieve complete disruption (Fig. 10), although trials with improved formulations in 1981 have reduced this level.

The pheromone of the rice stem borer, *Chilo suppressalis*, is a mixture of two aldehydes, (Z)-11-hexadecenal and (Z)-13-octa-decenal. Field-cage trials carried out in the Philippines at the International Rice Research Institute showed that the fully stabilised microencapsulated formulations of these aldehydes persisted for over 30 days and virtually eliminated the laying of fertile eggs during this period (6).

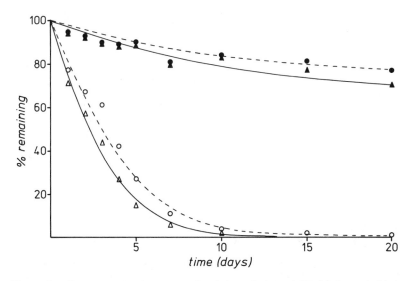

Figure 3. Exposure of microencapsulated formulation of Z9-14:Ac and 14:Ac containing N,N'-dimethyl UOP 688 in Egypt. Key: Z9-14:Ac, exposed (△) and shielded (▲); and 14:Ac, exposed (○) and shielded (●).

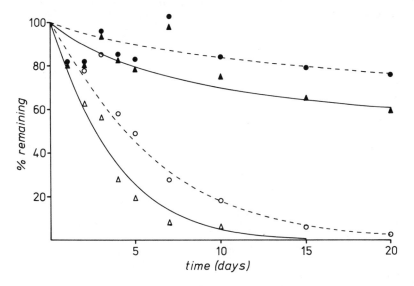

Figure 4. Exposure of microencapsulated formulation of Z9-14:Ac and 14:Ac containing Waxoline Black in Egypt. Key: Z9-14:Ac, exposed (△) and shielded (▲); and 14:Ac, exposed (○) and shielded (●).

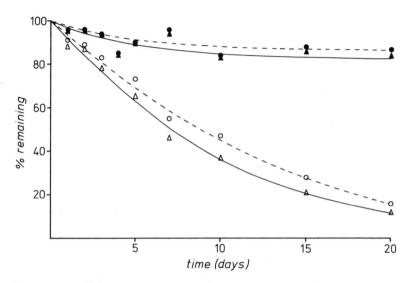

Figure 5. Exposure of microencapsulated formulation of Z9-14:Ac and 14:Ac containing N,N'-dimethyl UOP 688 and Waxoline Black in Egypt. Key: Z9-14:Ac, exposed (△) and shielded (▲); and 14:Ac, exposed (○) and shielded (●).

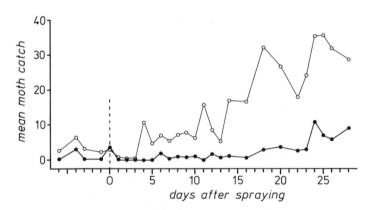

Figure 6. Communication disruption of Pectinophora gossypiella *in 0.01 ha plots in Egypt with microencapsulated pheromone applied at 5 g a.i./ha. Key: control (○) and treatment (●).*

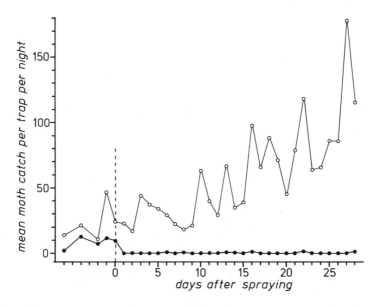

Figure 7.   Communication disruption of Pectinophora gossypiella *in 0.01 ha plots in Egypt with microencapsulated pheromone applied at 20 g a.i./ha.   Key: control (○) and treatment (●).*

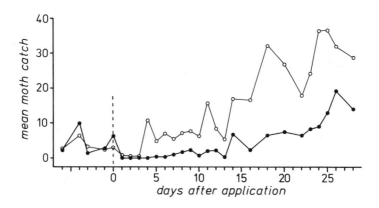

Figure 8.   Communication disruption of Pectinophora gossypiella *in 0.01 ha plots in Egypt with "NoMate PBW" hollow fibers applied at 4 fibers/m² (5.6 g a.i./ha). Key: control (○) and treatment (●).*

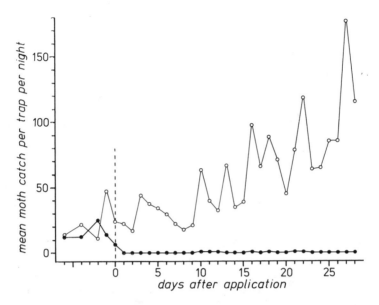

Figure 9. Communication disruption of Pectinophora gossypiella *in 0.01 ha plots in Egypt with "NoMate PBW" hollow fibers applied at 16 fibers/m² (22.4 g a.i./ha). Key: control (○) and treatment (●).*

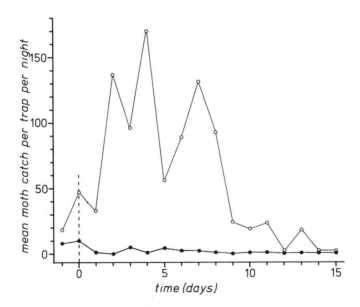

Figure 10. Communication disruption of Spodoptera littoralis *in 0.01 ha plots in Egypt with microencapsulated (Z,E)-9,11-tetradecadienyl acetate applied at 40 g a.i./ha. Key: control (○) and treatment (●).*

Acknowledgments

The authors are indebted to many people for their invaluable contributions to this work, notably Peter Beevor, Ralph Lester, Alan Cork and Maureen Eyles at TPI, Malcolm Alder at ICI and Philip Hunter-Jones and Lawrence McVeigh of COPR. The field work in Egypt was part of a collaborative project between the Overseas Development Administration and the Egyptian Academy of Sciences, and assistance by Ain Shams University and the Egyptian Ministry of Agriculture is gratefully acknowledged.

Literature Cited

1. Nesbitt, B.F.; Hall, D.R.; Lester, R.; Marrs, G.J. UK Patent 8007581.
2. Campion, D.G.; Lester, R.; Nesbitt, B.F. Pestic. Sci. 1978, 9, 434–440.
3. McVeigh, L.J.; Paton, E.M.; Hall, D.R. Proceedings 1979 British Crop Protection Conference - Pests and Diseases, 409–419.
4. Hollow fibre formulation "NoMate PBW" supplied by Controlled Release Division of Albany International Company.
5. Shani, A.; Klug, J.T. J. Chem. Ecol. 1980, 6, 875–881.
6. Dyck, V.A.; personal communication.

RECEIVED February 24, 1982.

Release Rates of Pheromones from Hollow Fibers

J. WEATHERSTON and M. A. GOLUB

Albany International, Controlled Release Division, Needham Heights, MA 02194

M. H. BENN

University of Calgary, Department of Chemistry, Calgary, Alberta, T2N 1N4, Canada

Following a review of methods used to measure the release rate of pheromones from hollow fiber formulations two newly designed pieces of apparatus are described. Results of release rate studies obtained with the new apparatus are discussed.

The increased use of pheromones to monitor and/or control insect pests is directly related to the development of efficacious controlled release formulations. In order for such formulations to be effective it is necessary to know the rate of release of the pheromone and the longevity of the formulation. It has been noted (1, 2) that the release rate of volatile materials appears to be related to the method used in the measurement hence studies were undertaken to improve the methodology by which release rates are measured.

Release Rate Methodology

Although a more detailed overview of release rate methodology has already been presented (3), a brief outline of the methods currently being used in this laboratory is in order here. Release rate methodology can generally be divided into two categories: those methods which measure the amount of material remaining in the release device, called residue analysis methods, and those methods which measure the amount of material being released, called effluent analysis methods.

Residue Analysis Methods. Among the methods routinely used for residue analysis are determination of weight loss, extraction of the residue and quantitation of the extract by either gas chromatography or liquid scintillation counting, and measurement of the increase in void space within a fiber with time, or meniscus regression method.

0097-6156/82/0190-0145$06.00/0

   While each of these methods has its own advantages and disadvantages all of the residue analysis methods are more rapid
than effluent analysis methods.

   Weight Loss. Of all the methods the weight loss method is
the most simple, requiring only the calculation of the difference
in weight as a function of time. However, the method also has
some serious disadvantages, the major one being that no qualitative information can be obtained about the active ingredient.
For example, changes in the composition of the active ingredient
caused either by degradation of the pheromone or a change in the
ratio of components of a multicomponent formulation would not be
detected by this method.
   In practice, two other disadvantages have also been observed. For very slow releasing components it has been found that
the decrease in weight over a reasonable time period can be too
small to detect on the balance required to accommodate the weight
of the release device itself. Secondly, adsorption or absorption
of materials from the atmosphere by the release device itself can
sometimes actually result in a weight gain with time.

   Meniscus Regression. The meniscus regression method is
comparable to the weight loss method in its simplicity and rapid-
ity. The increase in the length of the void space between the
open end of the hollow fiber and the meniscus of the column of
liquid is measured periodically using a Wilder Varibeam™ optical
comparator which has been described by Weatherston et. al. (3).
Since this instrument can also be used to measure the internal
diameter of the fiber, the volume and therefore the weight of
material lost can be calculated as a function of time.
   This method can more easily be used to detect the vaporiza-
tion of very slow releasing pheromones than the weight loss method since the optical comparator can detect changes in length of
± 0.002 mm which represents less than 0.1 µg for most pheromones
and fibers. The method is, however, limited to transparent release devices in which the release of pheromone can be directly
related to a change in the length or width of the space occupied
by the pheromone, and like the weight loss method, offers no qualitative information concerning the active ingredient.

   Extraction and Quantification of the Residue. When gas chromatography is used as the method of quantification, extraction
provides the most information of any of the residue analysis
methods since both qualitative and quantitative information can
be obtained. In this way any changes in ratios or the presence
of degradation products in the residue can also be detected.
   This method is, however, the most time consuming of the residue analysis methods. Other disadvantages are that the method is
destructive as once the pheromone has been extracted from the device, it is useless. Since examination of the residue at various
time periods is necessary to establish a release rate this neces-
sitates a large sample size.

Some of the disadvantages of each method separately can be partially overcome by employing a combination of methods. One such combination which has been used to good advantage in this laboratory has been that of meniscus regression to establish a release rate over a specified period in combination with extraction at more widely spaced intervals to determine the composition of the residue.

Major Disadvantages of Residue Analysis. In the foregoing discussion several advantages and disadvantages of the various methods have been discussed, but the most severe limitation of the residue analysis methods has not been touched upon. That disadvantage is that none of these methods provide any direct information about either the quality or quantity of the material actually released. If volatile degradation products are produced, this information would not be detected nor would the ratio of components actually released be directly measurable. Since the material released is the active ingredient of any controlled release system, this lack of information is a serious drawback to dependence on residue analysis for release rate determinations.

Effluent Analysis Methods

The effluent analysis methods commonly employed may be categorized in terms of air flow. One method measures the release of pheromone in a closed apparatus in which essentially no air movement occurs while the second method measures the release of pheromones into a stream of moving air. The first method has been called the static air method while the second has been called the moving air method.

While these methods both share the distinct advantage of looking directly at the active ingredient of the formulation they also share a number of disadvantages. Because of the small quantities released, sample preparation techniques can frequently be elaborate and therefore very time consuming. Since each step in the preparation of a sample is a potential source of error, this increased complexity can also decrease the accuracy of the method. Considerations of this type led this laboratory to the use of labeled pheromones to decrease sample handling and to increase the quantitative accuracy, however, liquid scintillation counting does not provide qualitative information about the labeled species.

Effluent analysis methods require the use of efficient inert trapping media to insure both quantitative trapping and quantitative recovery of the released materials. And, finally, effluent analysis methods do not give any information about the residue with the result that degradation of the formulation to nonvolatile components would be undetected. While this might not affect the amount of material in the atmosphere at a given time, it certainly would affect the longevity especially where the degradation was rapid.

Static Air. The static air method uses the more simple apparatus of the two methods. The method and apparatus originally used in these studies were adapted from those described by Baker et. al. (4). A container, usually a round bottom or recovery flask, equipped with a suitable stopper and holding device for a fiber was assembled and incubated for a specified time. As the pheromone was released from the fiber it recondensed on the walls of the flask from which it could be recovered by washing with a suitable solvent. The efficiency of glass as a trapping medium and the efficiency of recovery from the glass surface were established over a range of conditions (5). Quantitation was either by gas chromatography or liquid scintillation counting. The method has no special disadvantages other than those already discussed and the rate measurements seem to be unaffected by the size of the flask.

Moving Air Method. The moving air method is the method most generally favored in this laboratory since it most nearly simulates actual field conditions. The major disadvantage to this method is the possibility of breakthrough with very volatile pheromones. The phenomenon can frequently be controlled by variation of the collection period employed. In practice breakthrough has not been a significant problem with higher molecular weight pheromones even with collection periods of 24 hours.

Practical Considerations in the Development of New Methodology

It is logical to expect that when the release from a single controlled release device is measured by two or more methods the results will be comparable for a given set of conditions. The data listed in Table I were collected after measuring the release rate from hollow fibers first by either the static air or moving air methods and then by extracting and quantifying the residues from those same fibers. Not only is there no agreement between the rates determined by the two effluent methods, but there is also no agreement between the rates calculated by using two different methods on the same set of fibers.

Results such as those reported in Table I are clearly unacceptable. It was decided that the airflow method would ultimately be the most accurate and the most versatile so the attempts at improving the methodology began with a critical examination of the apparatus pictured in Figure I.

A close look at the entire apparatus revealed that the surface area of the sample chamber was in excess of 300 cm² whereas the surface area of 15 g of glass beads, approximately the amount that would usually be in a primary collector, was only 182 cm². When the fact that the released pheromone had to travel a distance of nearly 20 cm from the point of release to the point of entrap-

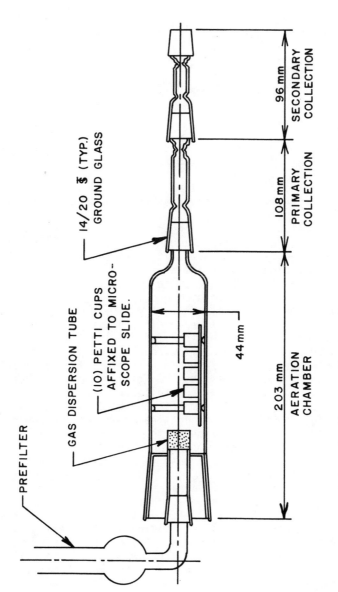

Figure 1. Air flow apparatus for measuring release rates of pheromones from hollow fibers.

Table I

Release Rates of (Acetyl-1-^{14}C)-Gossyplure from Hollow Fibers

Method	Mean rate* (μg/fiber/day)	Coefficient of Variability
Static air	0.34	2.3
Residue analysis (SA)	2.25	7.9
Airflow	0.74	0.7
Residue analysis (AF)	2.02	2.6

*Mean of at least 3 replicates

ment was taken into consideration, it became evident that there was a large area of glass available to trap the pheromone before it ever reached the collectors. The decision was therefore made to redesign the airflow apparatus on a smaller scale and thus improve its efficency. The redesigned unit is shown in Figure 2.

Mini-Airflow Apparatus. In contrast to the original apparatus, in which the aeration chamber was more than 20 cm long, the aeration chamber of the new unit is only slightly longer than the length of a single fiber and the entire apparatus, without the prefilter, is less than 12 cm in length. Glass beads are again used as the trapping medium and are held in place by a plug of silanized glass wool.

During operation the fiber or fibers are placed in the aeration chamber behind the prefilter and the luer joint is attached to a vacuum pump through a flow regulator. Air is drawn through the apparatus at a rate of 1 liter/min for specified time periods. At the end of the specified time the unit is detached from the pump and inverted over a suitable receiver. The accumulated pheromone is removed from both the beads and the aeration chamber by forcing solvent through the apparatus, in the direction opposite to the air movement, from a syringe attached to the luer joint. For ^{14}C labeled pheromones the apparatus is washed with 3x7 ml of hexane and each wash is collected in a separate scintillation vial.

Using this miniaturized apparatus the release rate of (acetyl-1-^{14}C)-gossyplure was measured from hollow fibers. A random sample of nine fibers was selected from which six were extracted to determine the initial loading and the remaining three used for release rate measurements. These three fibers were each placed separately in a mini-airflow unit and the daily release of (acetyl-1-^{14}C)-gossyplure measured for 13 days. To determine the reliability of the method these fibers were then extracted and the recovery calculated by adding the amount of residue in the fibers to the total amount released. These results are shown in Figure 3. The average recovery for the three fi-

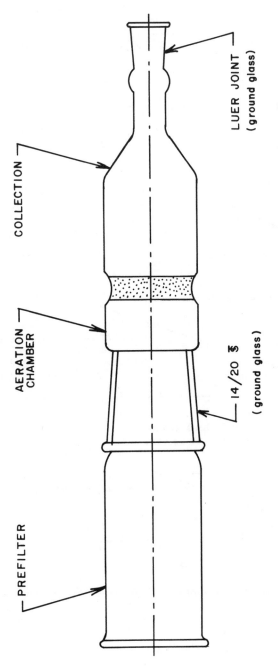

Figure 2.   Mini-air flow apparatus for measuring release rates of pheromones from hollow fibers.

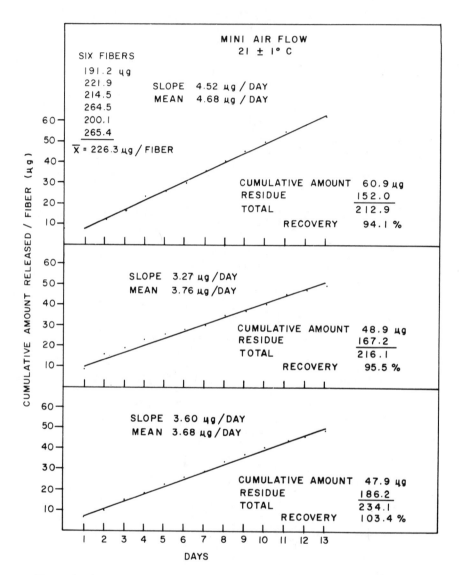

Figure 3. *Release rate profile of the release of (acetyl-1-14C)-gossyplure from single hollow fibers.*

bers is 97.7% (±5%). This ability to account for virtually all the pheromone originally present is of utmost importance since it indicates a quantitative recovery of pheromone from the apparatus and therefore an accurate measurement of the release rate.

To determine the versatility of the apparatus for measuring release rates in general, the release of (acetyl-1-^{14}C)-gossyplure was measured from rubber septa, a pheromone release device frequently used in field experiments. After 21 days of measurements, the septa were extracted and any residual activity in the apparatus was measured. The average recovery, detailed in Table II, was 97% indicating that the usefulness of the mini-airflow device was not limited to hollow fibers alone.

Table II

Recovery of (Acetyl-1-^{14}C)-Gossyplure
from Red Rubber Septa*

| | ug Recovered | |
	Unit 1	Unit 2
Cumulative release (21 days)	18.1	11.2
From apparatus	0.3	0.5
From septum	1049.4	1121.9
Total	1067.8	1133.6
Average initial loading	1133.3	
% Recovery	94	100

*Purchased from A.H. Thomas Co., Philadelphia, Pennsylvania

Recovery of Aldehydes. While the general usefulness of the new apparatus apparently could easily be extended to release devices other than hollow fibers, the same was not true for pheromones containing aldehydes. The primary problem in this regard was an erratic recovery of aldehydes from glass. When hexane solutions containing known amounts of aldehyde were applied to the walls of glass flasks similar to those used in the static air measurements and the stoppered flasks incubated overnight, the recovery of the aldehydes from either silanized or unsilanized flasks was low and erratic as shown in Table III.

However, when 1-^{14} C-(Z)-9-hexadecenal was applied to the same type of flask the recovery of radioactivity was 87% (±4%) from silanized and 88% (±3%) from unsilanized flasks. It, therefore, appears that although some reaction or reactions occur which alter the aldehydes so that they are, at present, not detected by gas chromatography, these same reactions do not prevent the reproducible recovery of the ^{14}C label. In addition,

when amounts of $1-{}^{14}C-(Z)-9$-hexadecenal were applied directly to
the walls of scintillation vials which were then incubated, the
recoveries were also high. The data shown in Table IV gives the
recovery in each of two 7 ml hexane washes, assayed spearately,
and the amount remaining on the vial walls. A change from hexane to toluene was subsequently made to increase the counting
efficiency. This change resulted in recoveries of 92% (+2%) and
95% (+2%) from silanized and unsilanized flasks respectively.

Table III

Recovery of Aldehydes Recovered from Glass Flasks

	% Recovered	
	Silanized[1]	Unsilanized
Aldehyde	Glass	Glass
(Z)-11-Hexadecenal	15(+12%)[2]	81(+16%)
(Z,Z)-11,13-Hexadecadienal	24(+70%)	63(+45%)
(Z)-9-Tetradecenal	21(+61%)	79(+15%)

[1]Glass was treated with hexamethyldisilazane.
[2]The numbers in parentheses are the coefficients of variability
of at least 7 recoveries.

Table IV

Recovery[1] of $1-{}^{14}C-(Z)-9$-Hexadecenal from Scintillation Vials

	% Recovered	
	Silanized[2]	Unsilanized
1st wash	69.2	78.7
2nd wash	17.8	6.8
Vial walls	2.0	2.1
Total	89.0(+5%)[3]	87.6(+2%)

[1]Mean of three replicates recovered with hexane.
[2]Vials were treated with hexamethyldisilazane.
[3]Coefficient of variability for total per cent recovered.

Mini-Static Air. These foregoing results indicated that the
use of a scintillation vial as a static air vessel might facil-
itate the measurement of release rates for compounds which were
difficult to recover from the more commonly used traps. With
this in mind, several orientations of fibers in scintillation
vials were experimentally tested until the one pictured in Figure 4 was finally adopted as being the most workable and repro-

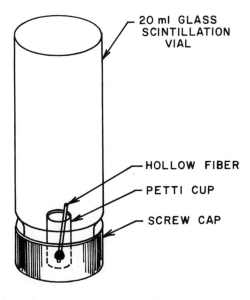

Figure 4. Mini-static air apparatus for measuring release rates of pheromones from hollow fibers.

ducible configuration. A series of such vials were assembled
and brought to thermal equilibrium. On a daily basis the fiber
was transferred to a new equilibrated vial and incubated. Scintillation cocktail was added to the old vial and the amount of
pheromone released during the previous incubation period was determined. The release rate of (acetyl-1-^{14}C)-gossyplure measured
in this manner from three replicates was found to be 2.27 μg/
fiber/day and the mean per cent recovery 106% based on an initial
assay of only 3 fibers.

Comparison of Mini-Airflow and Static Air Methods

The release rates of (acetyl-1-^{14}C)-gossyplure shown in
Figure 3 have a mean value of 3.80 μg/fiber/day. When the mini-
airflow rate is compared to the mini-static air rate of 2.27 μg/
fiber/day, it appears that the correlation of release rate measurements by different methods is no closer to solution than it
was before. Although the rates are still not in agreement,
there no longer is any question of achieving an accurate mass
balance at the end of a series of measurements by either method.
Since all the pheromone can be accounted for by both the mini-
airflow and the mini-static air methods, the differences appear
to be related to the conditions of the experiments.

The release of pheromone from hollow fibers as described by
Brooks (6) is dependent upon several factors including the movement of air past the open end of the fiber. In the mini-airflow
apparatus the open end of the fiber is in a stream of constantly
moving air so that any released pheromone is immediately swept
away from the end. In the static air apparatus the only air
movement across the open end will be as a result of diffusion.
Consequently the concentration of pheromone is more likely to
build up at the open end of a fiber in a static air apparatus
than at the open end of a fiber in an airflow apparatus with the
result that the release rate is lowered in the static air apparatus. Work is now in progress to evaluate the effect of air
speed on release rates.

The methodology developed to date has led to a marked improvement in the accuracy with which release rates can be measured from hollow fiber formulations. A careful combination of
methods such as the combined use of the airflow and extraction
methods discussed here with the appropriate combination of quantification methods will ultimately lead to a better understanding
of the various factors affecting the release of materials from
controlled release devices.

Literature Cited

1. Bierl, B.A.; DeVilbiss, D., Proceedings 1975 International Controlled Release Pesticide Symposium, 1975, p. 230.
2. Rothschild, G.H.L., Bull. Entomol. Res., 1979, 69, 115-127.
3. Weatherston, J.; Golub, M.A.; Brooks, T.W.; Huang, Y.Y.; Benn, M.H.; "Management of Insect Pests with Semiochemicals (1981)", Mitchell, E.R., Ed. Plenum Publishing Corporation, New York, 1981, p. 425.
4. Baker, T.C.; Carde, R.T.; Miller, J.R.; J. Chem. Ecol., 1980, 6, 749-758.
5. Golub, M.A.; Weatherston, J.; manuscript in preparation.
6. Brooks, T.W.; "Controlled Release Technologies: Methods, Theory and Applications, Vol. II", Kydonieus, A.F., Ed., CRC Press Inc., Boca Raton, Florida, 1980, p. 165.

RECEIVED February 24, 1982.

Evaluation of Controlled Release Laminate Dispensers for Pheromones of Several Insect Species

B. A. LEONHARDT

U.S. Department of Agriculture, Beltsville Agricultural Research Center, Beltsville, MD 20705

D. S. MORENO

U.S. Department of Agriculture, Boyden Entomology Laboratory, Riverside, CA 92521

Insect pheromones have been utilized in detection, surveying, monitoring and control programs with a wide variety of insect species. The nature of these biologically active compounds requires that they be formulated in controlled-release systems for prolonged biological effectiveness in field programs. The laminate dispenser system holds an inner reservoir layer containing pheromone between polymer membranes; the type and thickness of the membrane controls the rate of release. Such laminates have been investigated for the controlled release of the pheromones of corn earworm, tobacco budworm, gypsy moth, citrus mealybug, Comstock mealybug, and California red scale. Data on the release rates from various laminate dispensers are presented.

An intense interest in insect pheromones has been generated around the world in the past few years as applications have been developed to use these biologically active compounds to detect, survey, monitor populations and, in some cases, to control the target insects. Pheromones have been identified for ca. 250 insect species (1) most of which are Lepidoptera. Many of these compounds or blends of compounds have been used as baits in survey or monitoring traps to detect or estimate insect populations; some have also been successfully applied for mass trapping of males and as mating disruptants in insect control programs.

Lepidopteran pheromones generally contain 10-20 carbon atoms with a single oxygenated functional group such as acetate, alcohol or aldehyde; most have one or more sites of unsaturation. The compounds are quite volatile, and some, particularly the aldehydes, are unstable in air. All of these factors contribute to the need to formulate the pheromones in a

protective controlled-release system before they can be fully
and economically utilized in field programs for monitoring
populations or disrupting mating.

Over the years, many types of such controlled-release
formulations have been evaluated, such as cotton wicks, rubber
septa, polyethylene in various forms, cigarette filters, plastic
tablets, cork, wax, molecular sieve, etc. (2). In recent years,
commercial production of controlled-release formulations has
provided an economical and reproducible source of pheromone
dispensers and small particles for aerial dispersion. At the
present time, many of the insect pheromone programs involving
commercially produced, controlled-release products use one of
the formulations listed in Table I.

Table I
Commercial Formulations

Type	Manufacturer	Application	
		Dispensers	Sprayable
Gelatin Microcapsules	Eurand America (formerly National Cash Register Corp.) Dayton, OH		X
Polyurea Microcapsules	Imperial Chemical Industries, Ltd., Bracknell Berkshire England		X
	Stauffer Chemical Co. Richmond, CA		X
Hollow Plastic Fibers	Albany International Needham, MA	X	X
Polymer Laminates	Hercon Div. Health-Chem. Corp. New York, NY	X	X
Rubber Septa	Zoecon Corp., Palo Alto, CA	X	
Membrane/Reservoir	Bend Research, Inc. Bend, OR	X	

A number of the formulations containing pheromones have now been
registered by the Environmental Protection Agency for specific
use in insect management programs.

For controlled-release formulations to be effective and
economical, they must release most of their pheromone during the
flight season of the target insect. Too rapid a release would
necessitate costly retreatment, and too slow a release would be
wasteful of expensive pheromone by leaving a significant amount

of pheromone in the formulation after the insect season. The
development of a suitable formulation for a given compound or
group of compounds always requires a consideration of the
mechanism of release of the lure from the formulation and also a
determination of the rate at which the pheromone is emitted
under a given set of conditions.

The emission from a controlled-release formulation is
generally limited by a diffusion process which is controlled by
the concentration gradient across a barrier to free emission and
the parameters of the barrier itself (3). The rate of release
follows approximate zero order kinetics if the concentration
gradient remains constant; i.e., the rate is independent of the
amount of material remaining in the formulation except near
exhaustion. A large reservoir of pheromone is generally used to
attain a zero order release. Most formulations, however, tend
to follow first order kinetics, in which the rate of emission
depends on the amount of pheromone remaining. With first order
kinetics, $\ln [C_0/C]$ = kt where C_0 is the initial concentration
of pheromone, C is the residual pheromone content at time t, and
k is the rate of release. When C = $1/2\ C_0$, the half-life, $t_{1/2}$,
of the formulation is 0.693/k. Discussions of the theoretical
basis for release rates appear elsewhere (4-7)

The emission of a pheromone from a controlled-release
formulation can depend on the diffusion through holes in the
matrix or on the penetration of the compound through a wall or
membrane by absorption, solution and diffusion (8). Thus
variation in the parameters of the formulations, such as film
thickness, particle size, solvent, pore dimensions, etc., alters
the release rate. The design of the formulation must therefore
take into account the effect of each variable on the emission
rate in order to develop a system that is effective during the
appropriate cycle of the target insect.

Evaluation of a formulation design is best done under field
conditions where the impact of environmental conditions such as
sun, wind, rain and humidity can affect the release rate.
However, such field experiments are very costly and normally
cannot be used to screen a large number of formulations.
Therefore, laboratory methods have been developed to compare and
evaluate formulations under a fixed set of conditions. One
laboratory method involves aging of the formulations in a
chamber maintained at conditions that approximate those in the
field. Air passing through a chamber containing the dispenser
carries the emitted pheromone into an adsorbant trap for
subsequent analysis. In another method, the amount of pheromone
released from a formulation stored in a static air container at
a given temperature is determined. In a third method,
comparisons of residual pheromone in formulations are made by
weight-loss measurements. This procedure, however, suffers from
two significant drawbacks: 1) precise measurement of weight is
required, since typically the pheromone loss is in the mg range,

and 2) changes in weight caused by factors other than pheromone
loss (such as moisture absorption) must be minimized. Both of
these problems can be eliminated by solvent extraction of the
pheromone from the formulation followed by gas chromatographic
measurement of residual pheromone. A discussion of the variety
of methods for comparison of formulations is presented elsewhere
(2).

With the formulations described here (namely the laminate
dispensers), we had previously determined that the same pattern
of release was obtained by either of two methods of
measurement: collection of pheromone emitted under controlled
conditions of temperature and air speed (9, 10) or determination
of residual lure content by extraction and gas chromatographic
(GC) analysis. Since the latter method is simple and rapid and
provides reproducible data for the comparison of formulations,
it was used to obtain the data presented in this paper.

Formulations

Our Laboratory at the U.S. Department of Agriculture has
been involved in pheromone research for over 10 years and has
identified and synthesized such biologically active compounds
for many insect species. As a natural followup to those
identifications and syntheses, we have evaluated controlled-
release formulations of the pheromones for application in
monitoring and mating disruption programs. In early air-
permeation studies, we utilized a variety of substrates, such as
paper, cork, and molecular sieves (11) as matrices for
dispersion of disparlure in the gypsy moth, Lymantria dispar
(L.), program. Later, microcapsules, hollow fibers and laminate
flakes (11-16) containing pheromone (see Table I) were sprayed
in mating disruption programs for the gypsy moth and the
oriental fruit moth, Grapholita molesta (Busck). For monitoring
or survey traps containing pheromone as bait, the dispensers we
used initially were cotton wicks treated with the lure plus a
volatility extender (17), but we began using commercially
produced laminate dispensers (18, 19) as they became available
in the mid 1970's. Since that time, we have tested dispensers
containing pheromones for a variety of insects and have
determined their release rates (2, 9, 20, 21).

In this paper, I will describe results obtained in the
continuation of this work with the laminate formulations. The
laminates consist of an inner layer that contains the pheromone
plus laminating resins and outer layers made of polymeric
membranes. The lure diffuses from the inner reservoir through
the membrane layers and evaporates from the polymer surface. At
the pheromone concentrations that we have evaluated, the release
rate generally follows first order kinetics, with the rate
proportional to the concentration of lure. The parameters of
the polymeric membrane, including thickness and backbone

stiffness, and the nature of the pheromone itself (molecular
weight and polarity) control the rate of release (8). In
addition, we found that the size of the laminate flake or
dispenser also can alter the emission rate since some lure
escapes from the edges of the laminate as well as from the top
and bottom polymer layers (9).

Evaluation

The formulations were aged in a laboratory, a greenhouse,
or outdoors for 6 to 12 weeks. Periodically, samples were
removed and extracted with hexane or 20% acetone in hexane; the
extracts were analyzed by GC on a Hewlett Packard (Avondale, PA)
model 5830A instrument fitted with an auto sampler (model
7671A). The column was usually 1.8 m x 3 mm i.d., 3% SP2100 on
100/120 mesh Supelcoport (Supelco Inc., Bellefonte, PA); the
column temperature and program conditions varied with the
pheromone being measured. The measured lure contents were
expressed as percent based on the starting weight of the
dispenser or as mg per unit area of dispenser.

Heliothis Species

A number of insect species, including corn earworm,
Heliothis zea (Boddie), and tobacco budworm, Heliothis virescens
(F.), contain C_{14} and C_{16} aldehydes (C_{14}al and C_{16}al) in their
pheromone blends (22, 23). The release rate of a C_{14}al would be
expected to be higher than that of a C_{16}al because of the
molecular weight difference. A mixture containing 10%
n-tetradecanal (C_{14}°al) in (Z)-11-hexadecenal [(Z)-11-C_{16}^{1}al]
was formulated in four laminates, each made from a different
type of polymeric membrane as the top and bottom layers: vinyl,
Mylar^R-coated vinyl, acrylic, and rigid vinyl film. (The
saturated C_{14}°ol was used as a model because of availability.)
The total aldehyde content was approximately 2.3 mg per cm^2 of
laminate; 0.7% (of the pheromone weight) of 2,6-di-tert-butyl-4-
methylphenol (BHT) was added as an antioxidant. Table II gives
the thickness of the polymers.
The laminates were cut into 3 x 3 and 13 x 13 mm dispensers
which were then aged in 2 locations: within a greenhouse in
Maryland (mean temperature ca. 32°C) and outdoors in Texas
[daytime max. temp in low 20's (°C) and min. temp in low 10's
(°C)]. Some dispensers were coated with a film of RA-1645
adhesive (Monsanto Corp., St. Louis, MO) as used in aerial
applications, in order to determine the effect of the adhesive
on the release rate. At regular intervals, sample dispensers
were extracted (20% acetone in hexane) and the C_{14} and C_{16}
aldehyde contents were measured by GC. Figure 1 shows a typical
plot of % aldehyde vs days aged in the Maryland greenhouse. The
amount of lure released during the aging period was determined

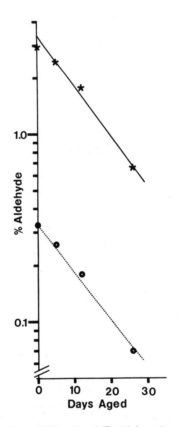

Figure 1.   Loss of tetradecanal (C₁₄al) and (Z)-11-hexadecenal (C₁₆al) from Mylar-coated vinyl laminate dispensers aged in a greenhouse. Key: C₁₄al (●) and C₁₆al (★).

from comparison of the amounts contained initially (at t_o) vs those remaining after aging. The half-lives, $t_{1/2}$, were calculated from the first order rate equation and these $t_{1/2}$ values for the dispensers are given in Table II.

Table II
Release of Tetradecanal (C_{14}) and (Z)-11-Hexadecenal (C_{16})
from Laminate Dispensers

| Polymer Film | | Dispenser Size | $t_{1/2}$ Values (days) | | | |
| | | | Greenhouse Aging | | Field Aging | |
Type	mil	mm	C_{14}al	C_{16}al	C_{14}al	C_{16}al
Vinyl	4	3 x 3	5	10	7	6
Vinyl	4	13 x 13	8	17	12	11
Rigid Vinyl	4	3 x 3	10	15	14	12
Rigid Vinyl	4	13 x 13	∿55	∿55	>60	∿55
Mylar[R]/Vinyl	0.5/2	3 x 3	12	14	6	6
Mylar[R]/Vinyl	0.5/2	13 x 13	39	31	11	10
Acrylic	6	3 x 3	18	25	18	15
Acrylic	6	13 x 13	>60	>60	>60	>60

These data illustrate several characteristics of the laminate dispensers. The smaller 3 x 3 mm dispensers release pheromone considerably faster than the 13 x 13 mm dispensers, due to the escape of lure from the edges or perimeter of the laminate; the ratio of edge to top and bottom surface area for the 3 x 3 mm dispensers is 12 mm edge:18 mm² area and for the 13 x 13 mm dispensers is 52 mm:338 mm². This effect has been reported earlier (9). The increased stiffness of the polymer backbone in rigid vinyl appears to reduce the emission of the aldehydes in comparison to the flexible vinyl of the same (4 mil) thickness. A Mylar coating significantly reduces the diffusion through the vinyl membrane, and the acrylic polymer is the least permeable of the 4 types tested. The release of both aldehydes was similar for all of the field-aged dispensers and for the greenhouse-aged Mylar-coated vinyl; on the other hand, the vinyl and rigid vinyl dispensers aged in the greenhouse released C_{14}al faster than C_{16}al. This is an important consideration when one is designing a formulation to release a constant ratio of a blend of pheromone components that differ in molecular weight.

Since aldehydes are readily oxidized in air to the corresponding acids, the extracts of the dispensers were also analyzed for C_{16} acid (GC using a 10% AT-1000 column at 250°C). No acid was detected in dispensers aged as long as 8 weeks; based on known quantities of C_{16} acid, less than 5%, if any, of the aldehyde originally present in the dispenser had oxidized to acid. Thus, the combination of polymer laminate plus BHT gave adequate protection of the aldehyde function. The coating of RA-1645 sticker had no significant effect on the release of the aldehydes.

In another experiment, a mixture of $C_{14}{}^{O}al$ (5%), (Z)-11-$C_{16}{}^{1}al$ (90%) and (Z)-11-$C_{16}{}^{1}ol$ [(Z)-11-hexadecen-1-ol] (5%) was incorporated into 3 other types of laminate dispensers at approximately 1.6 mg/cm^2; BHT was again added as an antioxidant. The alcohol was included since it is a component of the pheromone of the tobacco budworm. The laminates were cut into 4 x 20 mm dispensers and again were aged in 2 locations: at room temperature in a laboratory in Maryland and outdoors in Arizona (October-December). At selected intervals, sample dispensers from each location were extracted and GC analyses performed for all 3 components as well as for C_{16} acid formed through oxidation. Again, no C_{16} acid was detected in any of the formulations. The $t_{1/2}$ values for the laboratory and field aged dispensers are given in Table III.

Table III
Release of Tetradecanal, (Z)-11-Hexadecenal, and
(Z)-11-Hexadecen-1-ol

Polymer Film Type	mil	$t_{1/2}$ Values					
		Laboratory Aged			Field Aged		
		$C_{14}{}^{O}al$	$C_{16}{}^{1}al$	$C_{16}{}^{1}ol$	$C_{14}{}^{O}al$	$C_{16}{}^{1}al$	$C_{16}{}^{1}ol$
Acrylic	3	13	23	>30	6	10	∿17
Vinyl	2	3	12	13	3	6	8
Polymeric PVC	16	10	36	>>30	5	29	>30

The longer $t_{1/2}$ values for the laboratory-aged samples reflect the lower average temperature and air movement in the laboratory as compared to the field situation. As expected, the vinyl laminate released the aldehydes much too rapidly but could be suitable for the alcohol component. The acrylic laminate might also be useful for the alcohol. The polymeric PVC and acrylic laminates produced a greater difference in the release of the C_{14} and C_{16} aldehydes than did the Mylar-coated vinyl dispensers evaluated in the previous test. If the insect behavior requires that all of the components of the pheromone blend be released in a constant ratio and a rate proportional to their original concentration, then it is likely that the alcohol would have to be be formulated in one dispenser (e.g. in 3-mil acrylic) and the $C_{14}al$ and $C_{16}al$ in another (e.g. in Mylar-coated vinyl). Both of the dispensers could then be deployed at each location in the field.

Gypsy Moth

The gypsy moth sex attractant, disparlure [(cis)-7,8-epoxy-2-methyloctadecane], was one of the first pheromones to be evaluated in a laminate dispenser (19). Since 1975, all of the USDA survey traps for monitoring gypsy moth populations have utilized this type of dispenser. In the mid-1970's, 25 mg of racemic disparlure in a 25 x 25 mm dispenser (2-9 mil, vinyl polymer) served as the bait in the survey traps; however, once the more attractive synthetic, optically active (+)-disparlure became available in limited quantities, the survey dispensers were reduced in both size and lure content to conserve lure and to reduce the cost. Dispensers made from 2-mil vinyl film were evaluated in the following sizes and lure contents: 1 x 100 mm dispensers containing 0.12, 0.41, 0.90 and 1.70 mg of (+)-disparlure, 4 x 25 mm dispensers containing 0.15, 0.35, 0.86 and 1.76 mg, and 16 x 19 mm dispensers containing 0.46, 1.60 and 2.65 mg. The dispensers were aged outdoors in June and July, 1979 in Maryland and samples were evaluated after 0, 9 and 34 days of aging. The rate of loss of lure was independent of concentration within the laminate but did vary with dispenser shape. The 1 x 100 mm dispensers with maximum edge or perimeter relative to surface area lost 46% of the lure in the 34 days ($t_{1/2}$ = 38 days). The 16 x 19 mm dispensers with the least amount of edge relative to area lost only 33% ($t_{1/2}$ = 58 days), while the 4 x 25 mm dispensers gave an intermediate loss of 36% ($t_{1/2}$ = 53 days) over the same 34-day period.
On the basis of this test as well as the insect response to various dispenser baits, the 4 x 25 mm size dispensers were chosen for use in survey traps. The amount of synthetic (+)-disparlure available has dictated a concentration of 0.5 - 1.0 mg of lure per dispenser.

Disparlure has also been used in mating disruption experiments since the early 1970's and many of these experiments have demonstrated that mating of tethered females could be reduced by 90% or better in plots treated with controlled-release formulations of 20-50 g of racemic lure per hectare (12). One successful disparlure formulation for aerial application has been the laminate dispenser chopped into small flakes and coated with adhesive; a special apparatus has been described for the application of these flakes (see later chapters in this symposium by Plimmer et al. and Kydoneieus et al.). Based on the release data obtained with dispensers, the small flakes, which have a high amount of perimeter or edge relative to their surface area, whould be expected to lose lure at a faster rate than larger dispensers made of the same laminate.

Gypsy moth mating disruption tests in Maryland and New Jersey in 1979 utilized irregularly shaped flakes made of 3-mil vinyl film; the size of the flakes ranged from 5-40 mm ^{2}, with

an average size of about 9 mm^2. These laminate flakes were
slurried in RA-1645 adhesive and dispersed by aircraft. The
lure content was measured periodically in flakes recovered from
spray plots, and the $t_{1/2}$ was found to be about 59 days.

For 1980 tests in Maryland, an effort was made to release a
greater amount of the lure during the peak flight period of the
insect by reducing the size of the flakes. The laminate (again
3-mil vinyl film) was chopped into 2.4 x 2.4 mm squares (5.7
mm^2). The $t_{1/2}$, as measured from the lure contents of recovered
flakes, had only dropped to 53 days.

Since a still shorter $t_{1/2}$ was desired, the size of
the flakes was reduced to 1.9 mm^2 (0.8 x 2.4 mm) for the 1981
tests in Wisconsin and Michigan; the 3-mil vinyl film was again
used as the top and bottom layers of the laminate. The analysis
of recovered flakes gave $t_{1/2}$ values of 38 days in Wisconsin and
46 days in Michigan; this release seems ample for the program
needs, especially when one considers that the release from these
flakes would be somewhat faster at the warmer temperatures
encountered in Eastern states such as Maryland.

Citrus Insects

In 1980, we reported the structure of the Comstock
mealybug, Pseudococcus comstocki (Kuwana), pheromone as 2,6-
dimethyl-1,5-heptadien-3-ol acetate (24, 25) and in 1981, we
identified the pheromone of the citrus mealybug, Planococcus
citri (Risso), as (1R-cis)-(+)-2,2-dimethyl-3-(1-
methylethenyl)cyclobutanemethanol acetate (26). Both of these
pheromones have been synthesized and are currently being used as
baits in monitor traps in California and Texas. These compounds
are more volatile than the pheromones for most other insects and
therefore formulations for controlled release need to be
modified from those described earlier.

The optically active citrus mealybug (CIMB) pheromone was
intially formulated in laminate dispensers (13 x 13 mm) made of
9-mil vinyl and 16-mil polymeric PVC films. The dispensers,
containing 1 mg of pheromone, were aged outdoors in California
in late 1980 at temperatures ranging from 27-38°C in the day and
7-16°C overnight. Analysis for residual pheromone showed that
over 98% of the lure was released by the vinyl dispensers in the
first week ($t_{1/2}$ = 2 days), while the polymeric PVC formulations
gave a slower release with a $t_{1/2}$ of 11 days.

In a second test conducted to evaluate a dispenser that
would release lure more slowly, the CIMB pheromone was
formulated in both 16-mil polymeric PVC and 3-mil acrylic
films. The dispensers (50 x 12 mm), containing 10 mg of
pheromone, were aged in a greenhouse in California at
temperatures ranging from 21 to 36°C. Analyses for residual
lure content over a 7-week period gave $t_{1/2}$ values of 19 days
for the polymeric PVC and 44 days for the acrylic dispensers.

A final test to develop the best laminate dispenser for CIMB pheromone again compared the 3-mil acrylic vs 16-mil polymeric PVC dispensers containing 1 mg of pheromone. In this experiment, these samples were aged outdoors in California during the summer of 1981 (daytime temperatures of about 32-39°C and evening temperatures of 15-21°C). Rubber septa (5 x 9 mm, A.H. Thomas, Philadelphia, PA) impregnated with 1.2 mg of pheromone were also included in the evaluation. The analyses for residual lure of the 13 x 13 mm dispensers gave $t_{1/2}$ values of 42 days for the acrylic, 23 days for the polymer PVC and just 3 days for the rubber septa dispensers. These values are in very good agreement with the values obtained with greenhouse-aged dispensers.

A similar set of experiments was conducted for the development of a laminate dispenser for the Comstock mealybug (COMB) pheromone. Initially, 2-octyl acetate, which has a volatility and polarity similar to that of the pheromone, was used as a model and was incorporated in 3 types of laminate dispensers; 16-mil polymeric PVC, 6-mil acrylic and 9-mil vinyl films were used. Dispensers (13 x 13 mm) containing 2.7 mg of compound were aged in a greenhouse in California at temperatures ranging from 21-36°C and then analyzed. The 2-octyl acetate contents gave $t_{1/2}$ values of 1 day for the 9-mil vinyl, 8 days for the 16-mil polymeric PVC and greater than 40 days for the 6-mil acrylic film. On the basis of this experiment, the 9-mil vinyl and the 6-mil acrylic films were eliminated from consideration.

A second test compared 50 x 12 mm dispensers made of thinner acrylic film (3- instead of 6-mil) with those of 16-mil polymeric PVC; each dispenser contained about 10 mg of synthetic racemic pheromone. These were aged in a California greenhouse as in the previous test, and the lure contents were measured weekly for a 6-week period. The analyses showed similar $t_{1/2}$ values for the 2 formulations: 16 days for the 16-mil polymeric PVC and 18 days for the 3-mil acrylic dispensers.

The final dispenser evaluation with COMB pheromone again compared dispensers made of 3-mil acrylic film with those of 16-mil polymeric PVC; the 13 x 13 mm laminate dispensers each contained 1 mg of racemic pheromone. Also included in this test were rubber septa baited with 1 mg of compound. All of the dispensers were aged outdoors in California at temperatures ranging from 32-39°C in the daytime to 15-21°C overnight. Analyses of residual lure contents showed a $t_{1/2}$ of just 2 days for the rubber septa and approximately 15 days for the 16-mil polymeric PVC laminate; the 3-mil acrylic dispensers lost pheromone so slowly that $t_{1/2}$ could not be measured, but it was much greater than 70 days. On the basis of these tests, laminate dispensers made of 16-mil polymeric PVC were selected for the COMB monitoring program and have been successfully used in 1980 and 1981.

The California red scale, <u>Aonidiella aurantii</u> (Maskell),
is another important pest of citrus for which the pheromone has
been identified (<u>27</u>); although the pheromone consists of 2 novel
components: 3-methyl-6-isopropenyl-9-decen-1-yl acetate (I) and
(<u>Z</u>)-3-methyl-6-isopropenyl-3,9-decadien-1-yl acetate (II), only
compound I is used in the monitoring program for California red
scale (CSR) males. In an effort to evaluate laminate dispensers
for use in the CRS program, compound I was formulated in
dispensers made of 4- and 9-mil vinyl and 16-mil polymeric PVC
film. Dispensers (50 x 12 mm) containing approximately 10 mg of
I were aged in a California greenhouse (21-36°C) and analyzed
periodically for pheromone content. The $t_{1/2}$ values obtained
were 3 days for the 4-mil vinyl, 5 days for the 9-mil vinyl and
18 days for the 16-mil polymeric PVC films.

A second evaluation of dispensers was made to compare the
release of CRS pheromone component I from 9-mil vinyl and 16-mil
polymeric PVC laminates containing 1 mg in 13 x 13 mm
dispensers, following outdoor aging in California (32-39°C in
the daytime and 15-21°C overnight). Rubber septa baited with
0.1, 1.0 and 10 mg of I were also included in the test. As in
the previous test, the laminate of 9-mil vinyl film released the
pheromone very quickly ($t_{1/2}$ = 3 days) and was thus judged
unsuitable for controlled release in the CRS program. The $t_{1/2}$
values for the rubber septa (all 3 loadings) and the 16-mil
polymeric laminate were similar, approximately 40 days. The
analyses showed, however, that the rubber septa lost little, if
any, lure beyond the 40-50 day aging period; this was
particularly true of the lower loadings (0.1 and 1 mg) and
suggests that some of the pheromone is irreversibly "tied-up"
within the septa. This is not a problem with the polymeric PVC
laminate; pheromone continued to be released, even after 98 days
of field aging.

As in the case of the Comstock and citrus mealybug
pheromones, the release data for the California red scale
pheomone showed that the laminate formulation made with 16-mil
polymeric PVC film gave a long-lasting and suitable dispenser
for the monitoring program. The results of these dispenser
evaluations are summarized for citrus insect pheromones in Table
IV.

Summary

The evaluation of laminate dispensers based on lure
contents of aged samples has shown that the release rate for a
given pheromone can be widely altered by formulation in various
types of polymer membranes of differing thicknesses. The C_{14}
and C_{16} aldehyde pheromones are chemically stable in the
laminate, and a laminate can be selected for these aldehydes to
give a half-life, $t_{1/2}$, from 1 to 8 weeks, depending on the
duration of biological effectiveness needed. The C_{16} alcohol is

released much more slowly than the corresponding aldehydes. Small laminate flakes (0.8 x 2.4 mm) of 3-mil vinyl film give effective release of gypsy moth pheromone throughout the flight season in mating disruption experiments. The pheromones of three citrus pests, Comstock mealybug, citrus mealybug and California red scale, can each be formulated in 16-mil polymeric PVC laminate dispensers for effective use in monitoring programs. The data presented in this paper can be used as a guide to selecting appropriate laminate formulations of other insect pheromones.

Table IV
Release of Pheromones for Citrus Insects

Pheromone of[1]	Aging Conditions[2]	$t_{1/2}$ Values for Laminates of					
		4-mil Vinyl	9-mil Vinyl	3-mil Acrylic	6-mil Acrylic	16-mil PVC	Rubber Septa
CIMB	I		<2			11	
CIMB	II			44		19	
CIMB	III			42		23	3
COMB	IV		1		>40	8	
COMB	II			18		15	
COMB	III			>70		15	2
CRS	II	3	5			22	
CRS	III		3			∿40	∿40

1) CIMB = citrus mealybug; COMB = Comstock mealybug; CRS = California red scale.

2) Conditions: I = 13 x 13 mm dispensers with 1 mg of pheromone, aged outdoors (days 27-28°C; nights 7-16°C)
      II = 12 x 50 mm dispensers with 10 mg of pheromone, aged in greenhouse (21-36°C)
      III = 13 x 13 mm dispensers with 1 mg of pheromone, aged outdoors (days 32-39°C; nights 15-21°C)
      IV = 13 x 13 mm dispensers with 2.7 mg of 2-octyl acetate (not the pheromone), aged in greenhouse (21-36°C)

Literature Cited

1. Klassen, W. L.; Ridgway, R. L.; Inscoe, M. Chemical Attractants in Integrated Pest Management Programs, in "Insect Suppression Using Controlled Release Pheromone Systems", Kydonieus, A. F. and Beroza, M. Eds., CRC Press, Boca Raton, FL, in Press.
2. Bierl-Leonhardt, B. A. Release Rates from Formulations and Quality Control Methods in "Insect Suppression Using Controlled Release Pheromone Systems", Kydonieus, A. F. and Beroza, M. Eds., CRC Press, Boca Raton, FL, in Press.
3. Nightingale, W. H. Improved Means of Obtaining Sustained Uniform Emission of Bioactive Substances, in "Proc. 1979 British Crop Protection Conference-Pests and Diseases", 1971, 401-407.
4. Lewis, D. H.; Cowsar, D. R. Principles of Controlled Release Pesticides, in "Controlled Release Pesticides", Scher, Herbert B. Ed., ACS Symposium Series 53, American Chemical Society, Washington, DC, 1977, 1-16.
5. Lonsdale, H. F. Fundamentals of Controlled Release, in "Proceedings of the 1975 International Controlled Release Pesticide Symposium", Dayton, OH, 1975, 230-246.
6. McDonough, L. M. Insect sex pheromones: Importance and determination of half-life in evaluating formulations, Agricultural Research Results, ARR-W-1, U.S. Dept. of Agriculture, Yakima, WA, 1978.
7. McKibben, G. H.; Johnson, W. L. Bull. Entomol. Soc. Am. 1976, 22, 323-5.
8. Kydonieus, A. F. The Effect of Some Variables on the Controlled Release of Chemicals From Polymeric Membranes, in "Controlled Release Pesticides", Scher, Herbert B. Ed., ACS Symposium Series 53, American Chemical Society, Washington, DC, 1977, 152-167.
9. Bierl-Leonhardt, B. A.; DeVilbiss, E. D.; Plimmer, J. R. J. Econ. Entomol. 1979, 72, 319-21.
10. Beroza, M.; Bierl, B. A.; James, P.; DeVilbiss, E. D. J. Econ. Entomol. 1975, 68, 369-72.
11. Beroza, M.; Stevens, L. J.; Bierl, B. A.; Philips, F. M.; Tardif, J. G. R. Environ. Entomol. 1973, 2, 1051-7.
12. Schwalbe, C. P.; Paszek, E. C.; Webb, R. E.; Bierl-Leonhardt, B. A.; Plimmer, J. R.; McComb, C. W.; Dull, C. W. J. Econ. Entomol. 1979, 72, 322-6.
13. Gentry, C. R.; Beroza, M.; Blythe, J. L.; Bierl, B. A. J. Econ. Entomol. 1974, 67, 607-9.
14. Gentry, C. R.; Beroza, M.; Blythe, J. L.; Bierl, B. A. Environ. Entomol. 1975, 4, 822-4.
15. Gentry, C. R.; Bierl-Leonhardt, B. A.; Blythe, J. L.; Plimmer, J. R. J. Chem. Ecol. 1980, 6, 185-91.

16. Webb, R. E.; McComb, C. W.; Plimmer, J.R.; Bierl-Leonhardt, B. A.; Schwalbe, C. P.; Altman, R. M. Disruption Along the Leading Edge of the Infestation, in "The Gypsy Moth: Research Toward Integrated Pest Management", USDA Tech. Bull. 1584, in Press.

17. Beroza, Morton; Bierl, B. A.; Tardif, J. G. R.; Cook, D. A.; Paszek, E. C. J. Econ.Entomol. 1971, 64, 1499-1508.

18. Beroza, Morton; Paszek, E. C.; Mitchell, E. R.; Bierl, B. A.; McLaughlin, J. R.; Chambers, D. L. Environ. Entomol. 1974, 3, 926-28.

19. Beroza, Morton; Paszek, E. C., DeVilbiss, David; Bierl, B. A.; Tardif, J. G. R. Environ. Entomol. 1975, 4, 712-14.

20. Bierl, B. A.; DeVilbiss, E. D. Insect Sex Attractants in Controlled Release Formulations: Measurements and Applications, in "Proceedings of the 1975 International Controlled Release Pesticide Symposium", Dayton, OH, 1975, 230-46.

21. Bierl, B. A.; DeVilbiss, E. D.; Plimmer, J. R. Use of Pheromones in Insect Control Programs: Slow Release Formulations, in "Controlled Release Polymeric Formulations", Paul, D. R. and Harris, F. W., Eds., ACS Symposium Series 33, American Chemical Society, Washington, DC, 1976, 265-72.

22. Klun, J. A.; Plimmer, J. R.; Bierl-Leonhardt, B. A.; Sparks, A. N.; Primiani, M.; Chapman, O. L.; Lee, G. H.; Lepone, G. J. Chem. Ecol. 1980, 6, 165-75.

23. Klun, J. A.; Bierl-Leonhardt, B. A.; Plimmer, J. R.; Sparks, A. N.; Primiani, M.: Chapman, O. L.; Lepone, G.; Lee, G. H. J. Chem. Ecol. 1980, 6, 177-83.

24. Bierl-Leonhardt, B. A.; Moreno, D. S.; Schwarz, M.; Forster, H. S.; Plimmer, J. R.; DeVilbiss, E. D. Life Sciences 1980, 27, 399-402.

25. Negishi, T.; Uchida, M.; Tamaki, Y.; Mori, K.; Ishiwatari, T.; Asano, S.; Nakagawa, K. Appl. Ent. Zool. 1980, 15, 328-33.

26. Bierl-Leonhardt, B. A.; Moreno, D. S.; Schwarz, M.; Fargerlund, JoAnn; Plimmer, J. R. Tetrahedron Letters 1981, 22, 389-92.

27. Roelofs, W. L.; Gieselmann, M. J.; Cardé, A. M. ; Tashiro, H.; Moreno, D. S.; Henrick, C. A.; Anderson, R. J. Nature 1977, 267, 698-9.

RECEIVED March 1, 1982.

Formulations and Equipment for Large Volume Pheromone Applications by Aircraft

A. F. KYDONIEUS, J. M. GILLESPIE, M. W. BARRY, and J. WELCH—
Hercon Division, Health-Chem Corporation, New York, NY 10010

T. J. HENNEBERRY—U.S. Dept. of Agriculture, Western Cotton Research
Laboratory, Phoenix, AZ 85040

B. A. LEONHARDT—U.S. Dept. of Agriculture, Beltsville Agricultural
Research Laboratory, Beltsville, MD 20705

The U.S. Environmental Protection Agency has
issued Registrations and Experimental Use
Permits for seven Hercon controlled release
pheromone formulations of Disrupt (flake) for
population suppression by mating disruption of
pink bollworm, gypsy moth, western pineshoot
borer, peachtree borer, spruce budworm, artichoke
plume moth, and tobacco budworm. Application
rates range from 13 to 220 grams of flakes/acre
with 1.5 to 10 grams pheromone (active ingredient)
/acre, flake sizes from 1/32 to ¼ inch, and
duration of effectiveness from 2 weeks to 4
months. Special equipment was designed to apply
all of the flake sizes from aircraft at any rate
from 10 to 300 grams/acre. The equipment features
ease of application, 300-acre capacity/mission,
and quick clean-up and turnaround, all of which
are critical for commercial acceptance. Acrylic
emulsion stickers, compatible with the dispensing
machinery, were developed for use with the flake;
the stickers do not appreciably alter pheromone
release rates. Illustrations of mating disruption
are presented.

During the past several years, there have been numerous
reports describing the disruption of mating of important insect
pests by permeating the air with minute amounts of pheromone.
In this approach, pheromone-emitting particulates are broadcast
onto crops requiring protection from insects; the pheromone
enters the atmosphere above the crop, and insects find themselves unable to follow the pheromone trails that normally lead
them to a mate. The result: a sharp drop in mating, greatly
reduced reproduction, and fewer of the insect pests to contend
with in the future.

0097-6156/82/0190-0175$06.00/0
© 1982 American Chemical Society

The advantages of this technique, when properly used, are impressive. Pheromones are generally nontoxic and only small amounts per unit area are required. With pheromones, less insecticide is needed to manage insect pest populations, thereby decreasing insecticide costs and reducing insecticide residues to a minimum. Perhaps even more important, the decreased level of insecticide required to maintain control permits the beneficial insects to persist in greater numbers, allowing them to assist in restricting the proliferation of the targeted insect species.

There are other advantages, too. With lower levels of insecticide, wildlife are not apt to be affected and environmental problems associated with pesticide use, e.g., groundwater contamination and pesticide translocation, will definitely be diminished or eliminated.

The potential rewards of this approach have, in the past several years, prompted a number of companies to undertake the development of a commercial technology for applying the pheromone disruptant, thereby hastening the day when this important technique can be fully exploited as a safer and more efficient means of pest control.

Significant features of each company's system are 1) the method (or formulation) used to release the synthetic chemical disruptant into the atmosphere, and 2) the equipment and adjunct material for applying the pheromonal formulation to the crop requiring protection.

In this presentation, we describe the system developed by Hercon Division of Health-Chem Corporation, New York, New York 10010.

The Hercon Method for Controlling Pheromone Release

The Hercon dispenser is a patented 3-layer plastic laminate with the pheromone in the inner layer (1) (Figure 1). It can be fabricated in a variety of sizes and shapes; e.g., as small squares generally for use as the attractant in traps, or when used as a mating disruptant, as a flake (square or oblong) in sizes ranging from 1/32 to ¼ inch (Figure 2). In cases requiring greater rates of pheromone release, the laminate may even be ground to a powder.

When the dispensers are exposed, the pheromone gradually diffuses out through the outer layers of the laminate and is thereby slowly released to the atmosphere in a manner similar to that of an insect secreting its lure into the air to attract a mate. Since most pheromones are potent, very little pheromone is needed, and the dispensers can frequently contain enough pheromone to last an entire season. Location of the bulk of the pheromone in the inner layer of the plastic protects it from degradation by light, air and weather. This feature can be important because some pheromones are rather unstable (e.g., aldehydes), and they have to be protected until released,

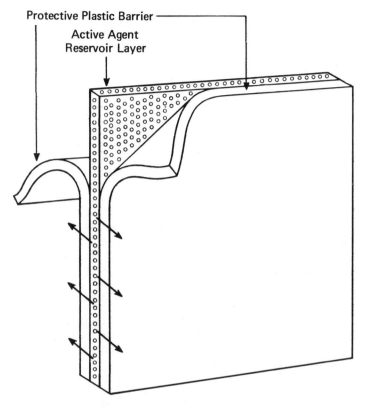

Figure 1.   The Hercon Disrupt pheromone dispenser.   Controlled amounts of active agent move from reservoir layer to surface and subsequent diffusion.

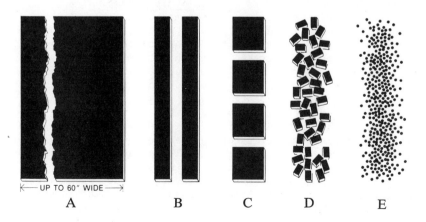

Figure 2.   Types of Hercon controlled release dispensers.   Key: A, sheet; B, strip or tape; C, wafer; D, confetti; and E, granules or powder.

especially since degradation products have been found to nullify
attraction of certain pheromones (2) (3).

A key feature of the Hercon dispensing system is the ease
of regulating the pheromone emission rate from the dispenser.
Thus, the emission rate may be adjusted by varying one or more
of a variety of parameters (4): 1) thickness of outer layers of
the dispensers, 2) pheromone concentration per unit area of the
dispenser, 3) size (area) of the dispenser, 4) amount of flake
applied per acre, 5) the polymer used to fabricate the dispenser,
and/or 6) polymer stiffness. Duration of effectiveness may also
be extended by increasing the thickness of the inner layer of
the dispenser, in effect increasing the size of the pheromone
reservoir of each flake.

In adjusting emission rates, the inherent volatility of a
chemical is taken into account. For example, a compound with
an 18-carbon chain will volatilize at a much lower rate than a
similar compound with a 12-carbon chain. Should we want to have
equal amounts of each of the compounds released per unit time,
we would have to make the outer layers of the dispenser much
thinner for the 18-carbon compound than for the more volatile
12-carbon one (other things being equal).

Hercon Insect Pheromone Mating-Disruptant Products

Hercon Division of Health-Chem Corporation has been awarded
registrations or experimental use permits by the U.S. Environmental Protection Agency for mating disruptant products for the
following insect species.

Artichoke Plume Moth	Platyptilia carduidactyla
Gypsy Moth	Lymantria dispar
Peachtree Borer	Synanthedon exitiosa
Pink Bollworm	Pectinophora gossypiella
Spruce Budworm	Choristoneura fumiferana
Tobacco Budworm	Heliothis virescens
Western Pineshoot Borer	Eucosma sonomana

Typical formulations and technology pertaining to mating-
disruptant products developed by Hercon for the above insects
are discussed below under the following three headings:

        Flake   (Disrupt Flake)
        Sticker   (Phero-tac Sticker)
        Dispensing Equipment

As has been noted in very early work on the technique of
mating disruption with pheromones (5), the technique works
best when applied before insect populations build up, e.g.,
early in the season, or with low-level populations. Use of
pheromone-mating disruptants directly against high insect
populations is not recommended. However, insect populations
that are already high are brought to low levels with insecticides before applying the pheromone-mating disruptant. Population levels may then be held low by continued use of the mating

disruptant. It is also noteworthy that pheromones are not useful in all instances; e.g., when there are too many pest species to control since each species would require a different pheromone. On the other hand, there are important large and small agricultural situations in which pheromones can have a very great impact. For example, pheromones can be used against cotton insects (6) (7), which require for control about 30% of all the insecticides used in the U. S., or they can be used as part of an integrated pest management program against a variety of other insect pests.

Disrupt Flake Parameters

Flake parameters are adjusted to provide the proper release rate for each individual species; i.e., one that will maintain a pheromone level in the atmosphere above the crop high enough to disrupt insect communication adequately for the desired time (usually for the duration of the insect mating flight or for an entire season. Multiple applications are frequently needed.). With temperature, sun, wind, rain, and other climatic factors being variable and influencing the pheromone release rate, and with the sensitivity of the target insect to its own pheromone being generally unknown, optimum emission rates for the dispensers cannot be predicted, and one must rely on the results of a series of formulations with a range of emission rates in small field trials to help select a suitable disruptant formulation. New formulations can then be compared with successful (effective) ones in the laboratory under standardized conditions to seek an improved product; e.g., one with a release rate equivalent to an effective formulation, but with a longer duration of effectiveness.

Table I presents sizes and weights (in terms of grams/acre) of Hercon's Disrupt flakes typically used against six insect pests.

Table I. Typical flake sizes and flake weights used against different insect species

Insect	Flake size	Flake weight (grams/acre)
Gypsy moth	1/32" x 3/32"	21
Pink bollworm	1/8" x 1/8"	60
Western pineshoot borer	1/8" x 1/8"	48
Peachtree borer	1/8" x 1/8"	14
Artichoke plume moth	¼" x ¼"	220
Tobacco budworm	¼" x ¼"	90

A comparison of several formulations in Table I shows how flake size and weight of disruptant/acre are varied to achieve

a desired pheromone emission rate. Thus, the gypsy moth pheromone with 19 carbon atoms in its molecule has a rather low volatility. To promote volatilization of the pheromone, the outer
layers of gypsy moth flakes are made thin (evidenced by less
grams/acre of flake being applied), and the flakes are small in
size, only 1/32 x 3/32 inch. (The smaller the flake, the more
edge of the laminate is exposed, and the greater the emission
rate.) The pheromone of the tobacco budworm moth with 14- and
16-carbon-atom molecules is much more volatile than the gypsy
moth pheromone. Therefore, its pheromone emission from the flake
is impeded more by fabricating tobacco budworm flakes with thick-
er outer layers and a larger size ($\frac{1}{4}$" square) than gypsy moth
flakes.

The data in Figure 3 illustrate another means of controlling release rates of flakes - by using different polymers.
Thus, the release rate of tobacco budworm pheromone was less
than half as much from an acrylic dispenser than from a vinyl
one (other things being equal).

Phero-tac Sticker

The application of the Disrupt flake onto crops can be
from the ground or from aircraft (with the latter probably
being the only large-scale practical approach). In either case,
it is necessary for the individual flake or other pheromone-
emitting particulate to adhere to the leaves of the crop upon
impact, as flakes falling on the ground are not effective, or
much less effective, in disrupting mating.

Hercon's Phero-tac sticker was developed to accomplish the
desired delivery of flakes onto the foliage with enough "green
tack" to cause each particle to stick.

The delivery operation requires the exercise of some judge-
ment. Flakes applied in a cool climate (e.g., at 20° C) will
lose less liquid carrier in passage through the air than flakes
applied in a hot climate (e.g., at 35° C). The difficulty in
hot climates is that the sticker on the flake may set (or dry)
completely before landing and not stick to the foliage. In cool
climates, the flakes must have enough time on the leaves for the
sticker to set (dry or cure completely and become water insoluble) before being subjected to rain (sticker in emulsion form
will be removed by rain). Thus, applications shortly before a
rain must be avoided to prevent washoff of flakes before the
sticker sets; excessive liquid in the emulsion is likewise not
helpful because it extends the setting time and the flakes may
fall off the leaves before the glue sets.

Hercon has circumvented much of the difficulty associated
with delivery of the flake through the use of newly and specially designed equipment and formulations. The equipment, to be
described later, covers the flake with enough of the sticker
liquid as it emerges from the flake hopper to retain its green

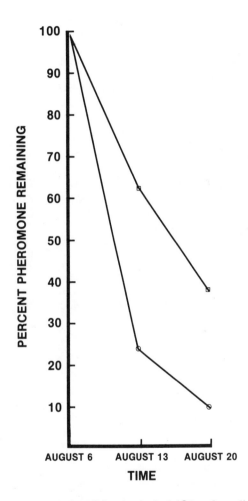

Figure 3. *Release rate of virelure flakes from vinyl (○) and acrylic (□) Hercon dispensers. Data show release rate of tobacco budworm pheromone.*

tack upon striking the foliage. This process is facilitated
through the use of an aqueous emulsion of the sticker. The
emulsion allows the water to flash evaporate in falling through
the air so the relatively involatile nonpolar sticker remains
on the flake and bonds with the leaf. The flake is mixed with
the sticker in the aqueous emulsion automatically at the time
of application. This practice avoids the difficulty of excessive leaching of pheromone from flake when flake and sticker are
mixed before application.

It is apparent that in coating the flake with sticker, volatility of the pheromone may be slowed. Undoubtedly, this occurs
to some extent, but it has not been found to be a problem with
the equipment used, presumably because of the thin film of
sticker applied to the flake. Table II presents data illustra-
ting this point with two different types of sticker, one a poly-
butene and the other an acrylic. Release rates of gossyplure
were found to be very similar to the rate with no sticker at all.

Table II. Effect of two stickers on release rate of gossyplure
from Hercon Disrupt Flake aged under laboratory conditions

| | $mg/inch^2$ remaining | | |
Day Exposed	No sticker	Polybutene	Acrylic
0	10.4	10.4	10.4
7	7.7	7.9	7.1
14	6.4	6.7	5.6
24	5.2	5.0	4.7

Additional features of Hercon's Phero-tac sticker that are
of interest are that the material is nontoxic to plants or
animals, non-corrosive, and it needs no cleaning after use.

Dispensing Equipment

The dispensing equipment used on aircraft is shown in
Figure 4.

The sticker liquid, which is in the tank shown below the
forward edge of the wing, is metered into a mixing chamber located immediately below the funnel-like structure located behind the tank. A wind-driven propeller mixes the sticker and
the flake, which originates in the funnel structure, and forces
the mixture out of the chamber via a screw-type arrangement toward a terminus equipped with a blade-like arrangement that
rotates rapidly in flight. With the aid of the slip stream, the
blade dispenses the flakes as individual particles.

Loading of sticker and flake is accomplished by installing
in place pre-packaged units of flake and sticker, making the
loading operation rapid and efficient. In saving application
time, this feature not only reduces the cost of treating by

Figure 4. Hercon flake-dispensing equipment used on aircraft.

aircraft, it allows the distribution of flake to be accomplished when time available for application is limited; e.g., by
weather.

Features of the dispensing equipment may be summarized as
follows:

>   Easy to install and use
>   Capacity of 300 acres
>   Lightweight
>   Night operation
>   Turnaround of 15 minutes
>   Quick cleanup
>   Good flake-sticker coverage

Figures 5 and 6 show typical performance of Hercon equipment in dispensing flake and sticker. Output in each case is
linear with the dial setting on the equipment, allowing very
close control of the amount of flake and sticker dispensed in
flight.

Mating Disruption Trials in the Field with Hercon flake

Examples of performance of Hercon flake in the field are
given below.

Figure 7 summarizes data on the release of the peachtree
borer pheromone, $(\underline{Z},\underline{Z})$-3,13-octadecadien-1-ol acetate, ODDA),
from a Hercon flake over a 70-day period. Figure 8 presents
similar data for release of gossyplure (1:1 $(\underline{Z},\underline{E})$- and $(\underline{Z},\underline{Z})$-7,
11-hexadecadien-1-ol acetate), the pink bollworm pheromone sex
attractant, from Hercon flake over a 3-week period. Emission
of pheromone is reasonably uniform in both instances.

Figure 9 shows the results of some 1979 trials in Arizona
in which pink bollworm trap catches were determined in cotton
fields treated with Hercon gossyplure flake and in "control"
fields; the control fields were actually fields treated with the
normal regimen of insecticides, i.e., conventional treatment.
Disruption of insect communication for mating is inferred from
the low level of insect captures in the gossyplure-treated areas
versus the insecticide-treated "control" areas, particularly
during the period after August 3. A more reliable estimate of
effectiveness is shown in Figure 10, which shows the damage to
the cotton (expressed as number of pink bollworm larvae per 50
bolls) in both the gossyplure-treated and the "control" plots
throughout the 1979 cotton-growing season. Damage in the pheromone-treated area is far less than in the "control" area. (Note:
Figure 10 includes a simultaneous experiment with tobacco budworm pheromone (virelure), which is not discussed here.)

Table III presents data obtained by exposing virgin clipped-
wing pink bollworm adult females on mating tables in the fields
treated with gossyplure at several levels.

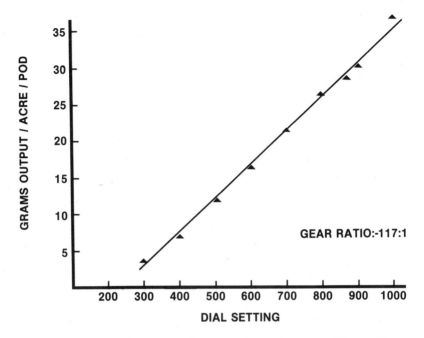

Figure 5. Typical performance of Hercon equipment in output of Hercon Disrupt flake.

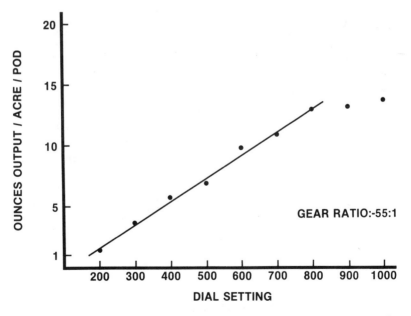

Figure 6. Typical performance of Hercon equipment in output of Phero-tac sticker.

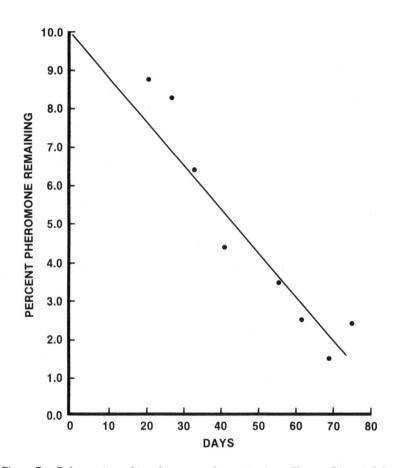

Figure 7.   Release of peachtree borer sex pheromone from Hercon Disrupt flake.

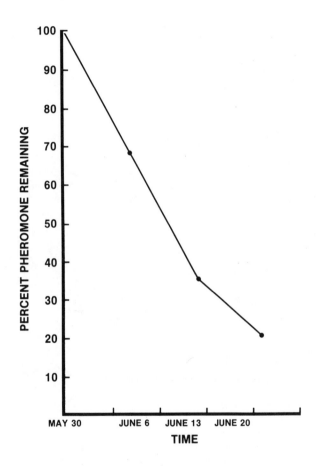

Figure 8. Release of pink bollworm sex pheromone from Hercon Disrupt flake. Key: vinyl, ●.

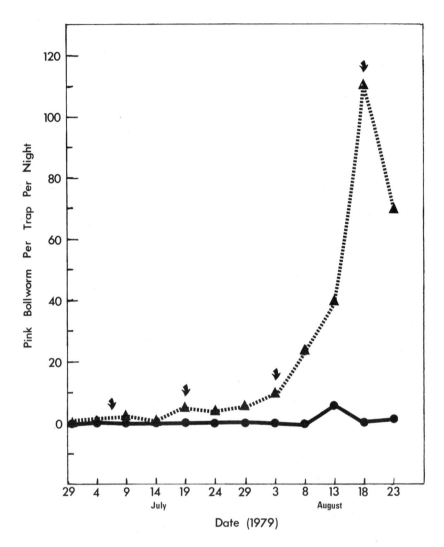

Figure 9. Trap catches in cotton fields treated with Hercon gossyplure flake and "control" plots (treated with normal regimen of insecticides). Key: gossyplure treated field (–●–); control field (· · ▲ · ·); and date of gossyplure treatment (↓).

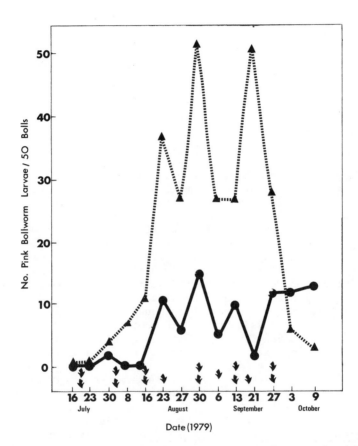

Figure 10. Damage to cotton by pink bollwarm in gossyplure-treated and "control" plots. Key: treated field (–●–); control field (· · ▲ · ·); gossyplure application (↓); and virelure application (↓).

Table III. Percent pink bollworm females mated in the field (40-acre plots) treated with gossyplure at several levels.

Treatment (grams a.i./acre)	Number of point sources/acre	Percent matings during night after treatment		
		2	13	17
PART I				
Control	--	25	49	57
1.0	1830	12	44	36
1.5	2750	0	3	6
2.0	12800	0	3	21
Part II		0	6	16
Control		68	79	35
1.2	2200	4	7	3
1.5	9600	16	33	12

Suppression in mating, shown in Part I of Table III, increased as the level of gossyplure increased to 1.5 gram a.i./acre. The decrease in mating suppression at the 2.0 gram compared to the 1.5 gram a.i./acre level on the 17th night may be ascribed to disruption of mating being more effective with the lesser number of point sources or the strength of the source. A repetition of the experiment (Part II, Table III) appears to indicate that the lesser number of point sources with higher pheromone content per point is the important feature contributing to the greater suppression of mating.

Similar experiments have been conducted in which insect damage or mating suppression have been demonstrated with other insect species. Table IV shows that 7.5 gram a.i./acre was highly effective in suppressing mating of the artichoke plume moth.

Table IV. Percent artichoke plume moths mated in control and pheromone-treated plots.

Pheromone treatment (grams a.i./acre)	Percent matings during night after treatment	
	1	7
Untreated	73	87
5	20	22
7.5	0	2

In a 200-acre trial in California conducted in 1979 with a Hercon formulation of the pheromone of the western pineshoot borer, Dr. Gary E. Daterman of the USDA Forest Service reported a 99.6% reduction in mating based on trap-catch data. Percent control achieved, based on infestations per 100 terminals, was found to be 88%.

A series of tests with gypsy moths at a number of locations in Northeastern U.S., have similarly shown that mating of this insect can be suppressed by pheromone treatment and that the treatment is especially effective in eradication or suppression of the insect in newly or sparsely infested areas. (See chapter by J. R. Plimmer et al.).

LITERATURE CITED

1. Beroza, M; Paszek, E. C.; Mitchell, E. R.; Bierl, B. A.; McLaughlin, J. R.; Chambers, D. L. Environ. Entomol. 1974, 3, 926-8.
2. Tumlinson, J. H.; Mitchell, E. R.; Browner, S. M.; Mayer, M. S.; Green, N.; Hines, R.; Lindquist, D. A. Environ. Entomol. 1972, 1, 354-8.
3. Beroza, M.; Gentry, C. R.; Blythe, J. L.; Muschik, G. M. J. Econ. Entomol. 1973, 66, 1307-11.
4. Kydonieus, A. F. in "Controlled Release Pesticides", ACS Symp. Ser. 53, Scher, H. B., Ed., American Chemical Society, Washington, D.C., 1977, chap. 14.
5. Beroza, M.; Knipling, E. F. Science 1972, 177, 19-27.
6. Henneberry, T. J.; Gillespie, J. M.; Bariola, L. A.; Flint, H. M.; Lingren, P. D.; Kydonieus, A. F. J. Econ. Entomol. 1981, 74, 376-81.
7. Brooks, T. W.; Doane, C. C.; Staten, R. T. in "Chemical Ecology: Odor Communication in Animals", Ritter, F. J. Ed., Elsevier/North Holland Biomedical Press, Amsterdam, 1979, 375-88.

RECEIVED March 1, 1982.

Field Measurements of Pheromone Vapor Distribution

ALAN W. TAYLOR

U.S. Dept. of Agriculture, Beltsville Agricultural Research Center, Agricultural Environmental Quality Institute, Beltsville, MD 20705

Field measurements of concentrations of disparlure
in air under woodland plots treated with three
different slow release formulations showed that
all released pheromone for about one month: concentrations decreased about 80% in the first five
days and 90-98% over 35 days. Between 75 and 85%
of the disparlure remained in the formulations
after 35 days even though release had became very
slow. Measurements with other formulations
containing tetradecenol formate applied to corn
showed these were more efficient but not
persistent enough to control Heliothis Zea for
more than one month. No satisfactory
measurements of concentrations in vapor plumes
from point sources were possible even though
these may be as effective as broadcasts. Further
field research is limited by sampling and
analysis techniques and the need for better
micrometeorological data.

The object of the work described here was to measure the amount and distribution of pheromone vapor in the air under woodland canopies after aerial applications of slow release formulations of disparlure, and the rate at which these declined with time. A limited number of experiments have also been done Z-9-tetradecenol formate (TDF), a mating inhibitor for the corn earworm moth (Heliothis Zea).

In all the experiments, direct measurements of the concentrations of the chemicals in the air in the treated plots were made by drawing known volumes of air through adsorbing samplers placed at various heights within the plots. Details of the treatments and sampling procedures varied somewhat from experiment to experiment and have been described elsewhere (1, 2, 3, 4, 5.): only the more relevant details will be summarized here. Since current analytical

procedures are not sensitive enough to detect aerial
concentrations produced by applications of these chemicals
at the rates recommended for population control, the rates
used in the experiments described here were up to 500
g.a.i./hectare, which is about 25 times that used in
recommended practice. It is however believed that the results
obtained can be directly extrapolated downward to the lower
rates.

Broadcast Applications of Disparlure To Woodland

In an experiment in woodland at Beltsville, Maryland, in
September 1979, three separate formulations were applied by
air to separate 4 hectare plots (200 m x 200 m). The trees
were between 15 and 20 meters high in a dense stand of
deciduous species with a few intermixed evergreens.
   The first plot received 500 g.a.i./h of displarlure as NCR
gelatin-walled microcapsules containing 2% ai. The
formulation, applied as an aqueous suspension, also contained
1% of sticker to aid adhesion of the formulation to foliage.
The second plot received 500 g./h. as Herculite Corporation
sprayable laminate flakes containing 9.1% ai. The flakes
consisted of two layers of vinyl, each 0.08 mm thick on both
sides of a central porous layer containing the disparlure:
the surface area of the flakes was between 7 and 35 mm^2 per
side. The same sticker as that in the microcapsules was used.
The third plot received 330 g.a.i./h as "Conrel" controlled
release hollow fibers containing nominally 11.5% ai.: a
suitable sticker was also incorporated in the formulation.
(Note that the use of trade or proprietary names here or
elsewhere does not constitute an endorsement by the USDA).
   Air samplers were mounted in the center of each plot at
0.3, 2.0, 5.0 and 10 meters above ground. Samples were taken
for consecutive 4 hour periods in overall 24 hour runs on a
series of separate days up to 34 days after application.
Samples of the laminates and hollow fibers were recovered from
the plots and analyzed for pheromone residues on several days.
   Persistence Aerial concentrations are plotted as a
function of time for all three formulations in Figure 1. Each
curve shows the decline in concentration calculated in terms
of the average all heights and sampling periods in each 24
hour day. Rapid initial declines were evident over the first
few days in both the microcapsule and fiber formulations.
After the tenth day, when average concentrations were between
5 and 10 ng/m^3, all declined steadily to between 0.4 and 2.0
ng/m^3 after about 30 days.
   Similar results, presented in Table I, were obtained in an
earlier experiment in which the same microcapsules were
applied at 250 g.a.i./h to a 20 m high canopy of deciduous

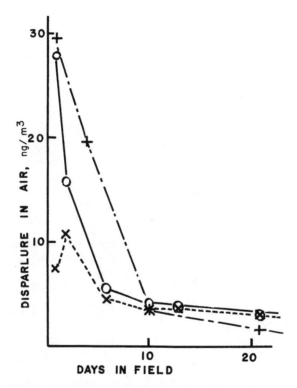

Figure 1.   Change in average concentration of disparlure in air under a deciduous woodland canopy treated in September 1979 with 3 slow release formulations: a) Conrel hollow fibers (— — —); b) NCR microcapsules (———); c) Hercon laminates (– – –) (adapted from 1).

Table I. Average Disparlure concentrations during
daylight hours at 0.3 and 8.0 meters above ground
under a woodland canopy treated in August 1976

Days after application	Concentration (ng/m^3) at height	
	0.3 m	8.0 m
1	76	50
2	46	25
3	29	16
7	22	4
14	2	1
22	3	1
34	2	0.5

woodland at Elk Neck State Park, Maryland in August, 1976 (2).
Although, as noted above, abnormally high application rates
were used in these experiments in order to ensure measurably
high concentration in air for the maximum time, the only
difference between these and applications at the normal
application rates of about 20 g. ai./h was that the number of
formulation particles was greatly increased. This suggests
that the data can be directly extrapolated downward to
estimate aerial concentrations in plots treated at recommended
rates, indicating that, 30 days after treatment, these should
be in the range between 0.10 and 0.02 ng/m^3 for the
microcapsules, flakes and fibers. These are about the same
order of magnitude as the reported threshold concentrations
for the response of many insects to pheromones (6). It is
therefore doubtful that the effectiveness of these
formulations is likely to be more than about 5 weeks under
conditions similar to those of these experiments. In warm
summer weather effective lifespans might well be less. It
should be noted that while this may be a sufficent period for
suppression of the mating of the gypsy moth which has a single
generation of insects each year, and which is sexually active
over a limited time period, single applications of these
formulations may be quite inadequate to produce pheromone
vapor concentrations over the longer time periods necessary to
control mating of other insect species with field populations
of a complex age structure. Persistence of formulations may
also be less where they are applied to shorter field crops
which have a more varied microclimate than in the rather
sheltered conditions under a woodland canopy.
     The effectiveness of the formulation in terms of the
amounts of material applied can be evaluated in terms of the

Table II. Effectiveness of slow release formulations
under field conditions.

Concentration In Air	Flakes	Fibers
	(ng/m^3)	
Initial	7.2	22.5
30 days	1.1	0.4
Decline	85%	98%
Disparlure Content of Formulation (%)		
Initial	7.1	5.0
30 days	6.0	3.8
Fraction released	16%	24%

data presented in Table II. These show that, even after the
pheromone emission from the particles had ceased, very large
fractions remained as unused and inactive residues. Data
suggesting a similar limitation on the effectiveness of the
same microcapsule formulation were obtained in an earlier
field experiment (7). The reasons for these declines in
emission from the formulations while they still contain large
fractions of the active ingredients are unclear, but they
indicate that detailed study of the weathering process and
their behavior in field conditions is necessary to improve the
effectiveness of their performance and reduce the amount of
active ingredient applied and the number of applications that
may be needed for long period insect behavior control.

Vapor Distribution The distribution of vapors with height
under the treated canopy is a complex function of the
distribution of the formulation over the ground, the
undergrowth and the canopy itself coupled with the mixing of
the air by wind ventilation and instabilities of temperature
gradients. Since however, turbulence is much less under the
canopy of a leafy wood than in an open field directly exposed
to wind and radiation, comparison of vapor profiles released
by different formulations during simular time periods will
give some indication of any differences in the distribution of
the sources, although these indications cannot be regarded as
quantitative. Such a comparison of the vapor profiles from
the three formulations used in the 1979 experiment are
presented in Table III. The most striking feature of these
results is the even distribution of the vapor with height for
all the formulations. As a broad generality the observation
appeared true throughout the 24 hour periods, no marked
concentration gradients being found at any time. In
comparisons between formulations the distribution of the
fibers was most uniform. The results for the laminates are

Table III. Distribution with height of average
disparlure vapor concentrations released from
three formulations applied to deciduous woodland
at Beltsville, Maryland in 1979.

Height (meters)	Formulation		
	Microcapsules	Flakes	Fibers
		(ng/m^3)	
10	14 a	5.2 ab	9.8 a
5	11 b	4.4 c	8.7 a
2	10 b	4.7 bc	8.7 a
0.3	9.4 b	5.5 a	10.6 a
Time periods averaged	26	27	25

Within formulations, means followed by the same letter are not
significantly different at the 20% level (by Duncans Multiple
Range Test).

---

ambiguous but do not indicate any tendency for accumulation at
any particular height. The data for the microcapsules
strongly suggests accumulation of the formulation on the
foliage of the canopy with somewhat smaller amounts on the
ground and undergrowth. These data contrast with those from
the 1976 experiment presented in Table IV. In these earlier
data gradients of concentration are clearly evident, both in
these results and others not presented here. These gradients
reflect the effect of the lower density of the canopy in the
1976 woodland plot in two ways. The higher concentrations at
the lower levels appear to reflect the greater penetration of
the formulation to lower levels in the thinner stand,
resulting in consistently higher vapor concentrations close to
the ground. Also the effect of increased wind ventilation in
the thinner canopy is clear in the data for the 15-17 hour
sampling period, where the higher release rate of the
microcapsules caused an increased concentration close to the
ground, while that at the 17 meter height fell due to
increased ventilation associated with greater afternoon wind
speed and turbulence.

In general these results suggest that the character of the
forest canopy will prove to be one of the most important
factors controlling the distribution of aerially applied slow
release formulations under woodland canopies and the ultimate
gradients of pheromone vapor under them.

Broadcast Applications of Tetradecenol Formate To Corn

The behavior of two other formulations, microcapsules and small plastic laminate flakes, containing tetradecenol formate, were compared in applictions to plots of mature corn at Beltsville, in August 1980. (8) The Z-9-tetradecen-1-ol formate (TDF) is a mating disruptant, rather than a true pheromone, of the Heliothis species of moths. It was selected in these experiments because reliable analytical methods were available (9) and its behavior was expected to be similar to that of the actual pheromones, whose chemical structures and properties are also similar. The two formulations were polyurea-polyamide microcapsules 5-microns in diameter supplied by ICI at Bracknell, Berkshire, England, and a small (3 mm side) plastic laminate formulation supplied by the Herculite Corporation of York, PA. The microcapsules were applied at 300 g of TDF per hectare and the flakes at 285 g/h. Both formulations were applied by air to mature corn 240-270 cm in height in clear, hot weather with a daily maximum temperature of 39°C.

Table IV. Concentrations of disparlure vapor between 0.3 and 17 meters height under a microcapsule-treated forest canopy at Elk Neck, Maryland, during daylight hours on August 18, 1976.

Time (EDT)	Concentration (ng/m^3) at height (m)		
	0.3	8.0	17.0
11-15	66	58	47
13-15	67	45	39
15-17	87	47	36
17-19	62	56	44

In order to measure the decrease in residues, samples of formulations were collected by hand immediately after spraying and at approximately weekly intervals. The laminates were collected by hand from plant surfaces scattered through the plot. For the microcapsules, 16 12.5 cm filter paper circles were laid together on a single plastic sheet to receive the spray. These were then clipped individually to plant leaves at 150 cm height at a number of locations throughout the plot. On each collection day, four filter papers were randomly collected and individually analyzed to measure the decrease in residues.

Concentrations of TDF in air were measured at five heights from 30 to 270 cm above the ground, using sampling and analytical methods described elsewhere (9). This sampling was done on days 3, 6, 10, 15, 21 and 31 after application: during each period of 24 hours samplers were changed at 2200, 0000, 0200, 0600 and 1200 EDT. Wind speed gradients, air temperatures and temperature gradients were measured continuously.

Persistence The results presented in Figure 2 show that the TDF residues disappeared from the laminates substantially faster than from the microcapsules. Expressed in terms of first-order kinetics the half-life in the capsules was 13.5 days and 5.3 days in the laminates. The latter figure agrees well with laboratory measurements of release rates: these also suggest that, in this formulation, the loss rate is controlled by diffusion outward to the edges of the flakes rather than through the exterior layers of vinyl covering.

The changes in average concentrations of TDF in the air under the crop canopy between 2200 and 0200 EDT for each sampling day are plotted in Figure 3. Since the stability of the air is greatest at this time, these values generally represent the highest levels of the diurnal cycle. They also co-incide with a critical time in the flight and mating behavior of the Heliothis moths. Both formulations gave the highest night-time concentrations after being exposed in the field for some time - the laminates after about 6 days and the microcapsules 15. It may perhaps be significant that both these correspond to the loss of about the same fraction of residues, at times slightly more than one residue half-life. The reason for the delay before the highest concentration, which was particularly marked for the microcapsule formulation, cannot be explained on the basis of changes in air temperatures or wind speeds. One possible mechanism could involve increases in permeability of the capsule walls and laminates as the formulations weathered, leading to increased specific rates of TDF release (i.e. flux/unit weight TDF). This would later be offset by the decreasing amount of residue present. As an alternative, the effect of adsorption and desorption by soil and plant leaf surfaces may be considered. Studies of the adsorption and desorption of insecticide and herbicide vapors have clearly shown that these are highly sensitive to water vapor and relative humidity. Similar organic molecules such as pheromones may perhaps be adsorbed during dry conditions in daylight hours and released by rising humidity and dew formation at night, thus tending to stabilize higher vapor concentrations under the still, moist canopy with stable air at night: it may be noted that such a mechanism will also present a naturally selective process favoring the mating of insects that release natural pheromones under such conditions. Despite these questions of mechanism it is clear

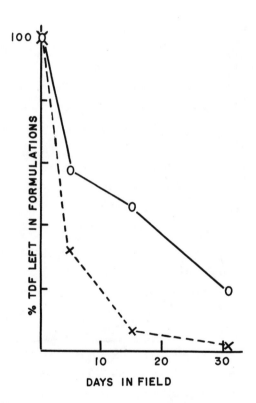

Figure 2. Decrease in residues of Z-9-tetradecenol formate in microcapsules (——) and laminated flakes (– – –) after field application to mature corn (8).

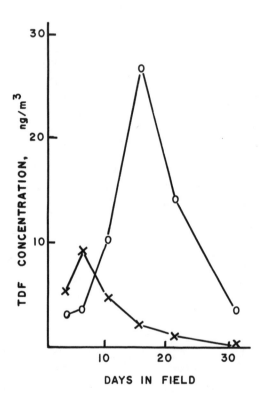

Figure 3.   Average concentrations of Z-9-tedradecenol-formate in air under mature corn between 2200 and 0200 EDT for 30 days after application of slow release laminate (×) and microcapsule (○) formulations (8).

that the plot that received microcapsules was bathed in
substantial concentrations of TDF vapor during the night hours
for about one month, whereas the laminate treated plot
contained only very low levels after about two weeks: this
difference appears primarily due to the pattern of release of
TDF from the formulations.

Effectiveness The two formulations used in the TDF
experiment differed from those in the earlier disparlure
experiment in that the release of the active chemical did not
cease until the residues were exhausted. If the steady
continuous release of chemical over a period of several weeks
is accepted as a criterion of effectiveness, these
formulations were clearly more effective than those in the
disparlure experiment. Where broadcast applications of
pheromones are made to control insect behavior in open, freely
ventilated agricultural crops, a very large fraction of the
pheromone released is lost by direct dissipation. Where, as
in the present case, the chemical is being applied to control
the behavior of night-flying insects this will be particularly
true because only that fraction released during night-time can
be counted as effective. Since no measurements of flux rates
were made no direct estimate of the amount lost is possible,
but indirect estimates of the amount required to permeate the
air during the insect flight period can be made. A simple
calculation shows that an average concentration of 10 ng/m^3
under the canopy of 1 hectare of crop with a height of 2.5 m
requires 250 micrograms. If the average wind speed is 1 m/sec
(2 mph) this must be replaced once per 100 seconds,
corresponding to 54 mg/night for the period 2100 to 0300
hours. This indicates that over a 30 day period about 1.62
grams or 0.5% of an application of 300 gr/hectare of active
ingredient can be regarded as contributing to insect mating
control. While the figures used in this calculation reflect
the artifical conditions used in the experiments described,
they illustrate a general principle and suggest that large
economies could be made in the amount of active ingredient
used by the development of formulations that release in
response to changes in humidity or temperature depending on
the release pattern desired.

Release From Point Sources

     In the light of the poor efficiency of broadcast
applications of pheromones in air permeation techniques the
alternative practice of scattering or placing a limited number
of release points at suitable distances apart over the treated
area merits consideration. Since each point will then act as
an individual source the treated area will be permeated by a
set of individual plumes of pheromone vapor whose direction
will vary with airflow through the crop or woodland. Large

concentration gradients may be expected from place to place within the stand.

Despite the difficulties of sampling and data interpretation two experimental attempts were made to measure the average concentrations produced in treatments of this kind and the way in which these declined with time. Neither produced meaningful data. It was clear that where individual sources each produce individual plumes with very high surrounding concentration gradients, sampling for two hour periods at points downwind from such sources produces time-averaged data that have little physical meaning other than indicating how long the source continues to give out a detectable plume.

The problem is well illustrated by measurements of TDF release from sources in a 10 meter spaced grid of individual points (Hercon plastic dispensers) in a mature corn canopy 180-200 cm high. Dispensers were at 108 and 180 cm heights. Samplers at 30, 100 and 150 cm height, 1 meter downwind from a source, detected time average concentrations of 50 ng/m^2 during the 2200-0200 hr period on the sixth day after exposure. This fell to 9 ng by the 27th day. No detectable amounts were found in similar samplers at 7 m equidistant from four separate grid sources. The data suggest a useful life of 30-40 days for plume emission by these dispensers.

A similar experiment to measure disparlure release from sources mounted on a 25 m grid under a woodland canopy, which had proved effective when evaluated by bioassay techniques measuring the mating of tethered females, gave a wholly ambiguous result in that concentrations were below detectable levels for the entire experiment.

The negative character of these results do not indicate that the use of a limited number of release points is biologically a less desirable practice, but rather that at the present time their evaluation by chemical methods requires sampling and analytical techniques beyond those currently available. This is indeed a point in their favor, indicating that, since they have been shown to be successful by biological techniques, this success must reflect the greater efficency with the smaller amounts of pheromone injected into the air. The use of a smaller number of source points also permits a greater degree of flexibility in the design of the formulations with more sophisticated release patterns.

General Conclusions

These experiments show that several formulations now available are satisfactory for the permeation of air volumes in woodland and crop canopies by pheromone vapors.

Some formulations cannot however be regarded as efficient sources in that they retain large fractions of the active chemical after release has ceased. This, coupled with the inherent inefficiencies of air permeation due to losses of chemical released continuously over periods when insects are inactive, means that the amount of chemical used to confuse insect mating behavior is a very small fraction of that applied. Improvement of this efficiency presents a major research challenge for the future.

Although intuition suggests that point source release techniques may be more effective, these methods are difficult to characterize by physical measurements because of the low sensitivity of our analytical methods and the difficulties of sampling release plumes. In the immediate future such methods can only be tested by direct measurements of effects on insect behavior.

We are also unable at the present time to sample and analyze many pheromone chemicals at any concentration - let alone the low biologically active levels - because of their instability upon the sampling media for the time they must remain there. Extension of our sampling and detection and analysis techniques to much lower levels presents another major research challenge. At the present time this is the main obstacle to progress. Beyond this point it is possible to foresee the development of a major research area in future which will contribute to large strides in the techniques of controlling insect populations and insect behavior.

In addition to the possible advances that will result from better chemical characterization of the direct effects of pheromone or other semiochemical treatments, contributions from other disciplines will be essential. An improved understanding of the micrometeorological conditions in which the insects live is essential both for an improved understanding of how these affect the insect behavior itself, including the use of the chemical communication systems they employ, and for improved understanding of the factors that affect the release patterns of the formulations used. Too often the behavior of the latter are evaluated and predicted in laboratory systems which grossly oversimplify the actual conditions of use. In practice, formulations are never used in isolation but rest upon soil or plant surfaces which are not passive but themselves respond to changes in temperature, radiation, moisture and air flow. These changes, apart from direct effects on the formulation are also likely to cause changes in the adsorption and desorption of the vapors themselves, whether they are emitted either by a calling insect or by an artificial formulation. In general, conditions under crop canopies at night have received little attention from agricultural climatologists who have concentrated primarily on the daytime effects of light, moisture and gas exchange because these are clearly of greater importance for the physiology of crop growth. As a

consequence, not only our understanding of the nature of
temperature, moisture and wind gradients within crops during
darkness is more limited, but the necessary instrumentation
for measurement of these in the more stable and less dynamic
conditions is lacking. The investigation of these subjects,
emphasizing the characterization of the micro-environment from
the point of view of the insect is likely to prove a highly
fertile research area.

   Two contributions from other disciplines are also
desirable. Improved use of pheromones and other semiochemicals will depend upon a much better understanding of the
biochemical and physiochemical mechanisms within the insect
that produces them and upon which they act. Such systems are
clearly complex and it seems highly probable that, once they
are understood, improved methods of inhibiting them will be
found, including the possible uses of individual inhibitory
chemicals or classes of them in place of the actual pheromones
themselves.

   Finally, improved techniques must be developed for
biological assay of the effectiveness of management techniques
for insect population control using semiochemicals. Since, in
the long run, the objective of all pest management techniques
is the control of populations, evaluation must be made using
population growth and decay as a basic statistic. Evaluation
based upon trap catches and mating rates in small confined
populations have uncertainties in the statistical significance
of the projections based upon them that are so large that
comparisons of the effectiveness of different treatments are
not possible at the level required for progress. The
development of improved techniques for measurements of
population sizes and changes and the statistics of their
interpretation probably represents one of the major challenges
not only in the techniques discussed here but in all other
approaches and contributions to integrated pest management
methods of insect control.

LITERATURE CITED

1. Caro, J. H.; Freeman, H. P.; Brower, D. L.;
   Bierl-Leonhardt, B. A. J. Chem. Ecology 1981, 7, 867-880.
2. Plimmer, J. R.; Caro. J. H.; Freeman, H. P. J. Econ.
   Entom. 1978, 71, 155-157.
3. Caro, J. H.; Glotfelty, D. E.; Freeman, H. P. J. Chem.
   Ecology 1980, 6 229-239.
4. Caro, J. H.; Bierl, B. A.; Freeman, H. P.; Sonnet, P. E.
   J. Agri. Food Chem. 1978, 26, 461-463.
5. Caro, J. H.; Freeman, H. P.; Bierl-Leonhardt, B. A.,
   J. Agri. Food Chem. 1979, 27, 1211-1215.
6. Caro, J. H. Chapter in "Insect suppression with
   controlled release pheromone systems" Kydonieus, A. K.
   and Beroza, M. (Eds) 1980. CRC Press, Boca Raton, FL.
7. Caro, J. H.; Bierl, B. A.; Freeman, H. P.; Glotfelty, D. E.;
   Turner, B. C. 1977. Environ. Entomol. 6, 877-881.
8. Caro, J. H.; personal communication.
9. Caro, J. H.; Freeman, H. P.; Bierl-Leonhardt, B. A.,
   J. Agric. Food Chem. 1979, 27, 1211-1215.

RECEIVED February 24, 1982.

Monitoring the Performance of Eastern Spruce Budworm Pheromone Formulations

C. J. WIESNER and P. J. SILK

New Brunswick Research and Productivity Council,
Fredericton, New Brunswick, E3B 5H1, Canada

New methods and apparatus are described for evaluating
the pheromone release characteristics of controlled
release formulations of Δ11-tetradecenal both in the
laboratory and following aerial application in the
field. Laboratory release rates determined by these
methods correlate well with rates observed in the
field.

The first attempts to influence mating behaviour of insects
in their natural habitat using broadcast formulations of sex
pheromones were conducted in the early seventies (1, 2, 3).
Beroza and his co-workers were able to show effective reduction
of mating success in low-level infestations of gypsy moth. In
the intervening ten years, a large number of "mating-disruption"
tests have been carried out against both agricultural and forest
insects (4). In virtually all instances, the problem has proven
to be far more complex than had been anticipated. The interplay
of biological, chemical and environmental factors has often led
to inconclusive field trial results. All too often the criteria
for success have not been met, while the cause for failure has
been obscured due to a lack of control or understanding of the
many variables.

One of these variables which obviously has a powerful influence on the result is the performance of the controlled release formulation. Happily, this variable is one which can, in
principle, be predicted and controlled.

The eastern spruce budworm, Choristoneura fumiferana, Clem.,
is one of the most economically important coniferous forest defo-
liators in the world. Several mating disruption experiments and
field trials have been conducted since the mid 1970's (5, 6). A
great deal has been learned in the process but no clear-cut conclusions regarding the feasibility of this approach as a management tool have, as yet, emerged. In an effort to simplify the
interpretation of future field experiments, we undertook a study

0097-6156/82/0190-0209$06.00/0

of the performance of various controlled release formulations of
Δ11-tetradecenal (TDAL), the sex pheromone of the eastern spruce
budworm. The aim was to find one or more formulations whose performance is predictable and efficient under our particular Northeastern environmental conditions. This paper describes and illustrates the techniques which were developed during the course of
that study.

Early experiments in our laboratory were concerned with
methods for sampling and analysis of TDAL from formulations (6),
insects (7) and from the forest atmosphere (8). This work was
largely founded upon concepts developed previously by Beroza et
al. (9, 10, 11). Since then, several other groups have applied
these concepts to the measurement of a number of different insect
pheromone release rates (12, 13). On the basis of our early
findings, we were convinced that the existing laboratory techniques for release rate determination from formulations were
inadequate. Laboratory tested formulations did not experience
the extremes of climatic variation which are the norm in the field
and consequently the release rate results were not transferable
to field performance.

In order to devise release rate methods which more closely
simulated the natural ageing process, we concentrated our efforts
on three main aspects of formulation evaluation: 1. Wind Tunnel
Ageing, 2. Effluvial Analysis, and 3. Atmospheric Concentration.

Wind Tunnel Ageing

Estimation of release rates by measurement of residual pheromone as a function of age is simple and economical. However,
since the method quantitates the amount of active ingredients
remaining in the formulation, the release rate is determined by
inference and does not take into account either degradation or
polymerization (14). Given a labile aldehydic pheromone such as
TDAL, these potential chemical changes must be taken into account.
Nevertheless, a laboratory method for simulating the natural
ageing process is essential.

Our first step was to develop a simple effective wind tunnel
which allowed us to age formulations in the laboratory under very
nearly natural conditions. Rather than attempt to control all
climatic parameters - temperature, pressure, humidity, light in-
tensity, wind speed, turbulence, etc. - which would have been a
formidable engineering task, a very simple design was chosen in
which only air speed, temperature and illumination were controlled.
The only real deficiency in this system is the lack of control of
humidity. However, that aspect was addressed separately by measuring the pheromone release rate of each formulation under conditions of very high and very low humidity. To date, none of the
candidates tested has shown a major humidity dependence.

The wind tunnel is shown in Figure 1. Air flow which is generated by an exhaust fan(D) can be varied from about 0.2 - 3.0

meters per second. Air is drawn through a heater(A), passes into
a mixing chamber(B) and into the test section(C). Access to the
test section is through a glass port(E) which also permits illumination by a standard sunlamp(F).

The operating conditions were arrived at empirically by adjusting temperature, wind speed and illumination on a diurnal
cycle until conditions were found which generated the same residual pheromone curve from a standard formulation in the tunnel
as that formulation experienced in the field. For this standard-
ization, one-eighth inch square Hercon flakes containing an average 13.3% TDAL by weight were used. This, as well as all other
formulations, were tested, as nearly as possible, in the same
form which they would have upon aerial application. They were
either coated or mixed with a recommended sticker and measured
aliquots were applied to rounds of filter paper. These were then
mounted on racks in the test section of the tunnel.

Having shown that identical ageing curves could be generated
with one formulation, the assumption was made that, under the same
operating conditions, other formulations would also behave
similarly in the tunnel and the forest. The operating conditions
of the tunnel are as follows:

> Light / 8 hours / 25 ± 1°C / 2.4 ± 0.1 m/s
> Dark / 16 hours / 14 ± 1°C / 0.55 ± 0.02 m/s

Effluvial Analysis

The actual air velocity in our original effluvial pumping
chamber (6) operating at 100 cc/min was about 5 cm/min, which we
regard as negligible. This, we believed, was largely responsible
for the lack of correlation between laboratory effluvial and field
residual pheromone release rates. In order to achieve realistic
air velocities, the diameter of the chamber was constricted considerably. The modified chamber is shown in Figure 2. The formulation is suspended on a wire in the case of large particles
while microdispersed materials are coated on a wooden popsicle
stick. These are placed in the inner tube(A) without touching the
glass wall. The tube is then inserted through a stopper into the
larger chamber(B) to within 2 cm of the bed of Porapak Q® resin(C)
supported on a glass sinter. The apparatus is placed in a temperature controlled chamber and connected to a nitrogen source. With
this arrangement, it is possible to generate air velocities up to
102 m/min in the inner tube. The samples are pumped for two
hours after which the resin and the walls of the main chamber are
washed with pentane. The TDAL is then quantitated by GC. Experiments run at maximum flow rates with two adsorbers in series,
showed negligible breakthrough (approx. 0.4%). The analytical
methods have been described in detail elsewhere (7).

This apparatus is used for two types of experiments:
1. Air velocity dependence of the release rate at 25°C and 2.7,

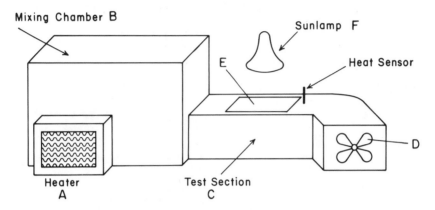

Figure 1. Wind tunnel for formulation aging. Dimensions of test section are: length, 40; width, 20; and height, 20 cm.

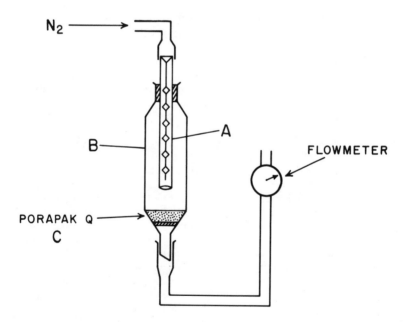

Figure 2. Modified effluvial pumping chamber. Key: A, inner tube; B, larger chamber; and C, Porapak Q resin.

26.9 and 102 m/min and 2. Temperature dependence of that rate at
four temperatures (10°, 15°, 20° and 30°C) and 26.9 m/min air velocity. All measurements are made on formulations which have been
aged for seven days in the wind tunnel.

Laboratory Release Rate Results

To date, over a dozen different formulations have been tested
in our laboratory. The following three examples represent a cross-
section of the test results and are presented to illustrate the
application of these new techniques: (A) Hercon 1/8" flakes,
(B) Capsular Products Company microcapsules, (C) Imperial
Chemical Industries microcapsules.

The wind tunnel residual pheromone curves for the three formulations are shown in Figure 3.

Examination of these curves suggests immediately that formulations A and C have reasonably constant rates of release and an
effective life-time of possibly forty days while B has lost 90% of
its active ingredient in the first week. The scatter about the
curve C may be due to the difficulty of reproducibly sampling
microdispersed formulations.

Figure 4. shows the effect of air velocity upon the release
rate of formulation "A" pumped for a total of six hours. Measurements were made at three flow rates (2.7, 26.9 and 102 m/min) and
at 25°C. Each point represents the average release rate of active
ingredient in the preceding two hours of pumping. At 2.7 m/min
the rate of release is constant with time of pumping, however, at
the higher flow rates about four hours of exposure were required
to achieve a steady-state rate of release.

A brief inspection of the slope of the residual pheromone
curve "A" in Figure 3. indicates a release rate of 4.3% per day on
day 7 and 1.1% per day on day 28. This translates to 2 µg/hr and
0.5 µg/hr, respectively. The seven day value compares very well
with 1.75 µg/hr, the steady state release rate measured at 102
m/min. (Figure 4.)

Figure 5. illustrates the steady-state release rate dependence of formulation C. In all cases, extrapolation of the curves
suggests that they plateau beyond about 100 m/min.

Finally, Figure 6. illustrates the temperature dependence of
formulation A.

Atmospheric Concentration Analysis

The final step in the testing program evaluates the performance of material aerially applied on small field plots. Since
this procedure aims to define the physico-chemical characteristics
of each product, not the biological effect, the following two
processes are monitored: 1. Residual pheromone from formulation
collected in the field zero to forty days post application and
2. atmospheric concentration of TDAL in the forest canopy.

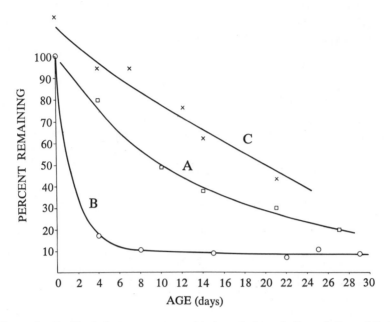

Figure 3.   Residual pheromone curves for formulations A, B, and C aged 3–4 weeks in the wind tunnel. Key: A, (□), Hercon flakes; B, (○), Capsular Products Company microcapsules; C, (×), Imperial Chemical Industries microcapsules.

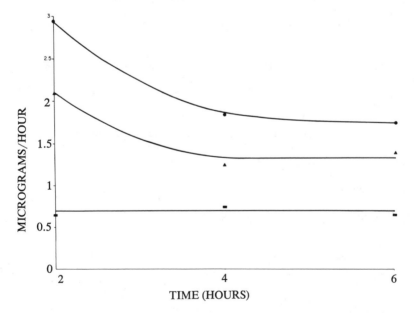

Figure 4.   Release rates at 3 wind velocities and 25°C of formulation A aged 7 days in the wind tunnel. Key: 2.7 m/min (■); 26.9 m/min (▲); and 102 m/min (●).

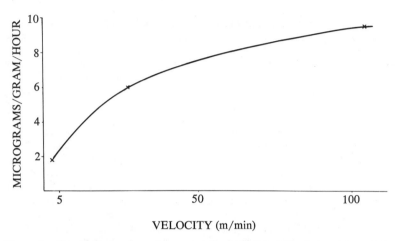

Figure 5. Wind velocity dependence of the steady-state release rate of formulation C at 25°C, aged 7 days in the wind tunnel.

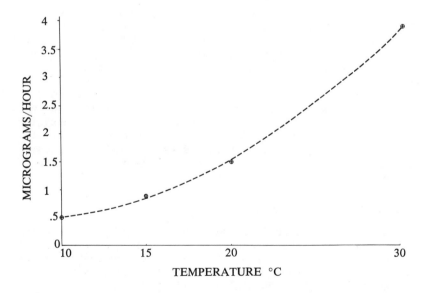

Figure 6. Temperature dependence of the release rate of formulation A aged 7 days in the wind tunnel at a wind velocity of 26.9 m/min.

The measurement of pheromone concentrations in the atmosphere was first described by Caro et al. for disparlure (11). Using an adaptation of his method we have developed sampling and analytical techniques specifically for tetradecenal quantitation.

The aldehyde is trapped with high volume samplers suspended in mid-canopy drawing 25 - 30 m^3/hr of air through a bed of 20 g of Amberlite XAD-2® resin. Originally the pheromone was derivatized to the pentafluorobenzyl oxime after solvent extraction from the resin (8). This was followed by silica gel column clean-up and quantitation by capillary GC with electron capture detection. This gave satisfactory results but was tedious and expensive. Also the derivatization produced electron capturing impurities which occasionally interfered with the tetradecenal peak. Our present method of analysis involves quantitation of TDAL using a capillary column GC/MS technique. The GC/Mass Spectrometer (Finnigan 4021) is used in the multiple ion detection (MID) mode under computer control. Specific ions in both the internal standard (decylbenzene) and the pheromone are recorded. The combination of high resolution capillary gas chromatography and the specificity of the mass spectrometer in the MID mode, makes this technique superior to the oxime method, albeit, a little less sensitive. (EC ∿100 pcg; MID ∿500 pcg)

Figure 7. shows the results of a field test of formulation A. The plastic laminated flakes were applied at a very high rate (500 g A.I./ha) on June 29th, 1980 to a mixed spruce/fir stand near Machias, Maine. The two curves represent mid-day (12:00 - 14:00) canopy concentrations of TDAL as well as the temperatures at that time. A period of heavy rain spanning July 1st and 2nd is reflected both in low noon temperatures and low aldehyde concentrations.

As mentioned above, the slope of residual pheromone curve "A" (Figure 3.) gives the release rate on day 7 as 2 µg/hr and on day 28 as 0.5 µg/hr, a ratio of 4:1. The atmospheric concentrations on those days were 5 ng/m^3 and 0.5 ng/m^3, a ratio of 10:1. In view of the unknown dilution effect of varying meteorological conditions, the relative correspondence between release rate and atmospheric concentration is quite good.

Conclusions

The techniques described and illustrated above now enable us to predict reliably the influence of the major climatic variables upon the rate of release of Δ11-tetradecenal from controlled release formulations under field conditions. With suitable recalibration, these methods should be applicable to any climatic conditions as well as any chemicals. A complete understanding of the release performance of a given formulation will, for the first time, permit us to interpret the results of a field treatment with the confidence that we are dealing with a biological effect, not a formulation effect.

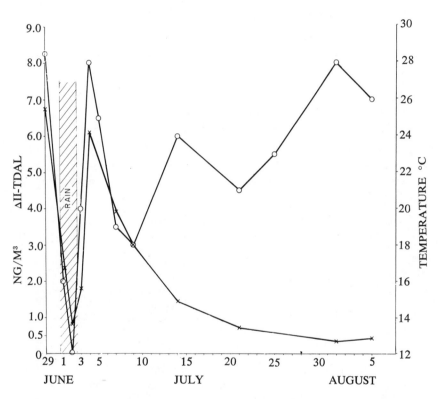

Figure 7.   Midday pheromone concentrations and temperatures at mid-canopy following aerial application of formulation A (500 g a.i./ha).

Acknowledgments

Support of this project by the Canadian Forestry Service, Environment Canada under Contract 1SS80-00066, is gratefully acknowledged.

Literature Cited

1. Stevens, L.J., and Beroza, M. J. Econ. Entomol. 1972, 65, 1090-5.
2. Beroza, M.; Stevens, L.J.; Bierl, B.A.; Philips, F.M.; Tardiff, J.G.R. Environ. Entomol. 1973, 2, 1051-7.
3. Beroza, M.; Hood, C.S.; Trefey, D.; Leonard, D.E.; Klassen, W.; Stevens, L.J. J. Econ. Entomol. 1974, 67, 659-64.
4. Roelofs, Wendell L., Ed. "Establishing Efficacy of Sex Attractants and Disruptants for Insect Control"; The Entomological Society of America, 1979.
5. Sanders, C.J. in "Chemical Ecology: Odour Communication in Animals"; F.J. Ritter, Ed.; Elsevier/North-Holland Biomedical Press: Amsterdam, 1979; p 281.
6. Wiesner, C.J., and Silk, P.J. "Report of Spruce Budworm Pheromone Trials - Maritimes 1978"; Maritimes For. Res. Centre, File Rep.; 62-64.
7. Silk, P.J.; Tan, S.H.; Wiesner, C.J.; Ross, R.J.; Lonergan, G.C. Environ. Entomol. 1980, 9, 640-644.
8. Wiesner, C.J.; Silk, P.J.; Tan, S.H.; Fullarton, S. Can. Ent. 1980, 112, 333-334.
9. Beroza, M.; Bierl, B.A.; James, R.; Devilbiss, D. J. Econ. Entomol. 1975, 68, 369-372.
10. Beroza, M. in "Pest Management with Insect Sex Attractants and other Behaviour Controlling Chemicals"; M. Beroza, Ed.; ACS Symposium Series, Number 23. American Chemical Society: Washington, D.C., 1976; p 99.
11. Caro, J.H.; Bierl, B.A.; Freeman, H.P.; Sonnet, P.E. J. Agric. Food Chem. 1978, 26, 461-463.
12. Look, M. J. Chem. Ecol. 1976, 2, 482-486.
13. Cross, J.H.; Tumlinson, J.H.; Heath, R.E.; Burnett, D.E. J. Chem. Ecol. 1980, 6, 759-770.
14. Cross, J.H. J. Chem. Ecol. 1980, 6, 789-795.

RECEIVED February 24, 1982.

The Utilization of Aggregation Pheromone for the Control of the Spruce Bark Beetle

ALF BAKKE

Norwegian Forest Research Institute, 1432 Ås-NLH, Norway

Drought and storm felling in the 1970's precipitated
an outbreak of *Ips typographus* that annually killed
millions of spruce trees in Norway. The identification and synthesis of the aggregation pheromone of
the beetle by 1977 raised the possibility of counter-
ing the continuing outbreak by manipulating the bark
beetles with their natural behavior-controlling chemicals. Pheromone dispensers containing a mixture of
2-methyl-3-buten-2-ol (1500 mg), (S)-*cis*-verbenol (70
mg) and ipsdienol (15 mg) were used as bait in drain-
pipe traps in a control program. About 600,000 traps
were deployed in South-Norway and about 40,000 forest
owners participated. Average catch per trap was 4700
beetles in 1979 and 7400 beetles in 1980. Total capture was estimated to be 2.9 billion beetles in 1979
and 4.5 billion beetles in 1980. Damage was significantly reduced in forest around the traps, particularly in lightly damaged areas. The beetles killed
about 10 million trees in Norway during the years
1978-80.

Bark beetles constitute a large group of insects which inhabit
forests all over the world. Most of the species are dependent on
dying trees or broken branches for their reproduction. Only a
few species are able to attack and kill healthy, living trees.
There are two main reasons for the extraordinary capabilities of
such aggressive bark beetles. Firstly: they have developed an
effective chemical signal system to coordinate the attack and to
aggregate in masses on selected trees. Secondly: they are supplied with patogenic blue stain fungi that invade the sapwood of
the tree, induce water stress and render the tree susceptible to
colonization by the beetles.

The development of chemical technology in recent years and
close cooperation between chemists and entomologists have pro-

0097-6156/82/0190-0219$06.00/0

vided the basis for identification, synthesis and utilization of
insect behavioral chemicals. Most of the fundamental research
work has been conducted in the United States with insects from
that part of the world, but some progress has also been made in
Europe. This paper describes a serious bark beetle problem in
Norway and how results of a research program on pheromones have
been utilized in efforts to deal with the problem.

The bark beetle *Ips typographus* is the main pest of mature
spruce in most of the spruce forests of Europe and Northern Asia.
Under epidemic conditions, the beetles are able to overcome and
utilize relatively healthy and vigorous trees. Control measures
have been undertaken in Europe for more than 200 years, mainly by
harvesting or burning infested trees and by felling trap trees.
A trap tree is a living, mature spruce which is felled in spring
and left in the forest during the main flight period of the
beetles. The trunk is being colonized by the beetles and then removed from the forest while the beetles are still in the bark.
Felling trap trees and removing infested trees in summer while
the brood still was in the trees, were the only control methods
used in Scandinavia until 1979. These methods are still used in
many countries in Europe, but they are expensive and have many
disadvantages.

Control by attempting to remove infested trees from the forest is particularly difficult because only 4-6 weeks are available from the time the infested trees can be identified by their
yellowing crowns until the broods begin to emerge. In a country
such as Norway with large forest areas and difficult, often steep
terrain, the removal of infested trees in such a short period is
often impractical. The need for more effective and economic control measures was urgent.

A research program led by the author of this paper was therefore established for the period 1974-78 with cooperation between
entomologists at the Norwegian Forest Research Institute and chemists at the Chemical Institute at the University of Oslo.   The
chemical research was led by Professor Lars Skattebøl. The aims
were to identify the aggregation pheromone components of *Ips typographus*, to develop syntheses for commercial production and to
work out methods for the utilization of pheromones in beetle control. The control measures presented in this paper are mainly
based on results from this research program.

Pheromone Components

The major aggregation pheromone components of *I. typographus* are
cis-verbenol and 2-methyl-3-buten-ol (1). *Cis*-verbenol is common
to several *Ips* species (2). Only the (S)-(-) enantiomer evokes
response in field experiments (3). Methylbutenol is specific to
I. typographus. A different isomer is part of the pheromone of
the related species *Ips cembrae*, which lives on larch in Europe
(4). Ipsdienol (2-methyl-6-methylene-2, 7-octadiene-4-ol), which

is shared by most *Ips* species, also occurs in *I. typographus*, but
seems to play a minor role in the aggregation pheromone complex.
The addition of ipsdienol increases the number of beetles captured in traps by an average of 50%, mainly by increasing the number of females. (R)-(-)-ipsdienol is identified from the species
(5), and field data indicate that this enantiomer is more attractive than its antipode. Ipsenol (2-methyl-6-methylene-7-octene-
4-ol) and verbenone, two other components of several bark beetle
species, have also been identified from the gut of *I. typographus*
male beetles. Both of these components inhibit the response to
the aggregation pheromone (6). They are released after the females have entered the gallery and seem to regulate the gallery density and cause the shift of attack to new bark areas or neighbour-
ing trees. Male beetles are the main producers of all pheromone
components, but only ipsdienol and ipsenol are male specific. Addition of host-volatile components to the ternary pheromone mixture
has not given an increased response in the field.

Application of Pheromones

A major outbreak of *I. typographus* in Scandinavia gave urgency to
the immediate application of the new knowledge of the beetles'
chemical communication system. Millions of spruce trees were killed by the beetle in the late 1970's and several million more
trees were overmature and highly susceptible to attack. This
great loss resulted from a combination of drought, storm fellings
and a high initial beetle population. Forests of more than a hundred thousand individual ownerships were infested in Norway. Attacks were scattered over an area of 140,000 km^2, which is equal to
the area of East-Germany or the state of Kentucky in the U.S.A. .
More than 90% of the ownerships are less than 100 hectares and be-
long to farmers whose main income is from farming. The severe and
widespread outbreak required active involvement of the government,
and an information campaign was started to convince the forest owners of the necessity of participation in a control program (7).
Pheromone dispensers were an important part of the control program.

The Pheromone Dispensers and their use. Two kinds of dispensers were used, both marketed by Borregaard Ind. Ltd., Sarpsborg,
Norway. 1. A plastic bag formulation produced by Celamerck, In-
gelheim, West-Germany. 2. A laminated plastic-tape formulation
produced by Herculite Products Inc., New York, N.Y., U.S.A. One
dispenser contained 1500 mg methylbutenol, 70 mg (S)-*cis*-verbenol
and 10 mg (1979) or 15 mg (1980) ipsdienol.
The pheromone release rate is insignificant at temperatures
below 10°C. Dispensers exposed inside traps in the first part of
May collected beetles during a two-month period, showing that they
were releasing pheromones during the beetles main flight period.

The dispensers were evaluated by the Pesticides Board of the Ministry of Agriculture and were approved for practical use in forestry.

The dispensers were applied as part of three trapping methods. 1. To aggregate the beetle in standing trap-trees which were felled and removed soon after attack (8, 9). 2. On lindane-sprayed log sections the beetles were killed by the insecticide after they had landed and started boring in the bark (10). 3. In baited traps (11). For economic reasons, the trap method was the one most commonly used.

The Traps. The trap (Figure 1) was made of polyethylene tubes with about 900 holes just large enough (3-4 mm) to allow beetle-passage. The tubes were capped at the top and had a funnel at the base leading to a collecting container. The model used in 1979 (Figure 1A) has a 135 cm long and 12.5-cm-wide pipe, and a bottle serving as the collecting container. Another model (Figure 1B) was introduced in 1980 on a minor scale. Its pipe is 130 cm long and 16 cm wide and there is a funnel at the base reaching 10 cm out from the pipe to collect beetles falling down along the pipe. The collecting container is perforated to permit rain water to pass through.

The 1980 trap model caught 50-100% more beetles than did the 1979 model, but an unsuccessfully made collecting container caused the escape of many beetles.

The price of one trap is about $5 and one dispenser costs about $2 when ordered in large quantities. The trap is easy to handle and can be used for several years.

About 600,000 traps were used in Norway in 1979 and 1980. They were positioned mainly in areas where trees had been killed by the beetle the years before. Guidelines were worked out for the location of traps.

Principles for the integrated Control Program. It must be emphasized that the use of pheromone dispensers and traps was only one part of the extensive control program. The major longterm objective was to stimulate increased harvesting of overmature stands in areas threatened by beetle attack. Government funds were available for the support of road construction in areas containing overmature stands, logging operations in steep terrains, and the purchase of modern logging equipment. A forest practices law was amended to prohibit storage of unbarked logs in the forest during summer, and to require clean-up after storm damage and logging.

In addition to mass trapping, felling and removal of beetle-infested trees was the main short-term measure. It was strongly recommended that infested trees should be removed from the forest in June-July before the adult of the next generation had emerged. Traps were to be deployed preferably in areas with minor infestation, one trap for every 3-5 trees killed by the beetle the previous year. Trapping was not recommended for saving old stands with extensive beetle infestation or stands severely weakened by

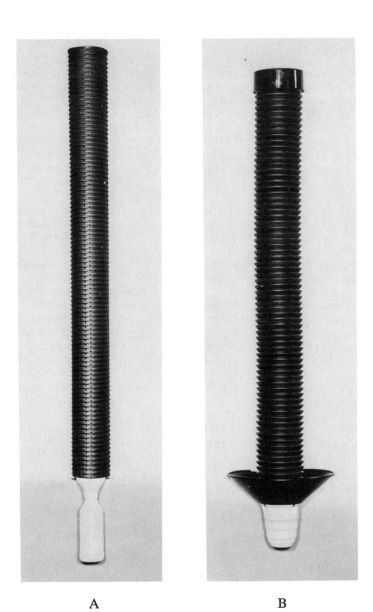

A   B

*Figure 1.   Pipe-trap models used in the bark beetle control program.   Key: A, the
1979 model, and B, the 1980 model.*

drought. Clear-felling was recommended for such stands. Instruc-
tions were given to place traps in cut-over areas to collect emerging beetles and thereby prevent their dispersal to other areas.

Results from the Mass Trapping Program. Traps located
throughout the beetle-infested areas were randomly selected for
analysis of beetle catches and the degree of changes in damage to
the forest around the traps. About 1% of the total traps were selected in 1979 (5679 traps) and ½% (2847 traps) in 1980. The
traps were emptied four times on given dates by local forest offi-
cers. A form was completed for each trap containing information
pertaining to forest condition within a radius of 100 m (1980),
number of beetle-killed trees each year, and the size of the catch
(12).

Trap Catches. Average catches per trap varied widely between
counties, the highest numbers being recorded in counties with the
most severe beetle attack (Table I). The catches were highest in
1980, when the average catch per trap was 7406 beetles. Average
trap catches in 1980 were 10,000-12,000 beetles in counties with
extensive beetle damage, whereas a catch of 2000-4000 beetles was
usual in counties with minor damage. Numbers of beetles caught
per trap varied from a few hundred to several thousand. Traps
located in areas cut-over the previous winter had the highest
catches. Average catches per trap decreased with increasing age
of the felling (Figure 2). Traps surrounded (within 100 m distance) by many beetle-killed trees from the year before, had the
highest catches (Table II). The total capture in Norway by all
traps was estimated to be 2.9 billion beetles in 1979 and 4.5
billion beetles in 1980.

Table I. Average catches per trap in some counties
of Norway. Number of test traps in parentheses.

	1979		1980	
Counties with severe damages				
Vestfold	6,919	(808)	11,701	(447)
Telemark	5,654	(1416)	9,967	(741)
Buskerud	4,682	(879)	7,806	(440)
Counties with minor damages				
Oppland	2,809	(490)	2,468	(253)
Hedmark	3,167	(596)	3,165	(206)
Østfold	3,634	(204)	4,337	(199)
All counties	4,701	(5679)	7,406	(2847)

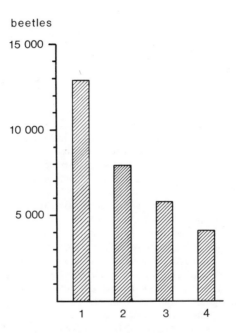

Figure 2. Average catches in 1980 in traps located in clear-cut areas of different ages. Key: 1, felling in 1979/80; 2, felling in 1978/79; 3, older felling; and 4, other open areas.

Table II. Average catches in 1980 from traps at
sites with different number of killed trees in 1979.
The damage is recorded within 100 m from the traps.
Number of traps in parentheses.

Number of killed trees in 1979.

	None	1-5	6-30	More than 30
All sites	3421 (438)	4342 (552)	7329 (1164)	12,150 (720)
Counties with severe damage Vestfold and Telemark	4712 (108)	5248 (189)	10,352 (504)	15,938 (345)

The main reason for the higher catches in 1980 was probably
the better location of the traps. In 1980, the guidelines emphasized the importance of setting the traps close to the ground, in
groups of 3-8 traps (about 5 m apart), in felling areas and at a
minimum distance of 30 m from mature spruce stands.

Trapping of non-target insects. More than 99% of the insects trapped were beetles of the genus *Ips*. *I. typographus* pre-
dominated, but a few per cent in certain localities were *I. dup-
licatus* (Sahlberg), which responds to the ipsdienol. The preda-
tory ant-like checkered beetles (Cleridae) *Thanasimus formicarius*
(L.) and *T. femoralis* (Zett.) are attracted by ipsdienol and *cis*-
verbenol (13) and were found in most traps. In 1979, the average
capture of *Thanasimus* was 1.6 per thousand *Ips*, and in 1980 1.4
per thousand. In the trap model of 1980, there were only 0.5 cle-
rids caught per thousand *Ips*. The chalcid wasp *Tomicobia seitnei*
(Ruschka), which oviposit in adult beetles while they are on the
surface of trees being colonized, were observed on the surface of
the drain pipe, but very few, if any, entered the trap.   These
data indicate that the drain pipe trap is very selective and better than sticky traps and window traps in preventing the killing
of useful predators.

Beetle-killed Trees around the Traps. In 1980, tree mortality was recorded within a distance of 100 m from the traps. The
trap sites were classified in four groups according to the number
of trees killed in 1979; none, 1-5, 6-30, and more than 30. There
was a clear correlation between the probability of new infestations around the traps and the number of beetle-killed trees the
year before traps were set up. At 70% of the sites with 1-5 killed trees the year before, no new infestations were recorded, and

Table III. Per cent of trap-sites where no new infested trees were recorded in 1980 in relation to
the number of beetle-killed trees in the same area
in 1979. Records are made within 100 m from traps.

Number of beetle-killed trees in 1979	All counties
None	95
1-5	70
6-30	51
More than 30	30

at 30% of the sites with more than 30 beetle-killed trees there
were also no new infestations (Table III).

The degree of change in the number of beetle-killed trees
from 1979 to 1980 was also recorded. There were distinct improvements on sites with all grades of damage in 1979. The changes
were about the same in most of the counties (Table IV). The total
site number showed improvement in 69% of the sites. 26% had un-
changed situation while 5% had increased damages.

Traps compared to Trap Trees. Trap-trees have been used in
Europe for more than 200 years to control the spruce bark beetle
(14). One reason for developing pheromone traps is to replace
the trap-tree method, which has become more and more expensive. An
average trap-tree (20 cm d.b.h.) will be occupied by about 6500
beetles (12). Average trap-catches in 1980 were 7406 beetles.
Within 2-3 weeks parent beetles occupying a trap-tree have utilized the phloem suitable for breeding so most of them emerge to
attack another tree. After 3-4 weeks only 10-20% of the beetles
are left. Trap-trees therefore have to be removed from the forest

Table IV. Degree of change in the amount of beetle-
killed trees around the traps in 1980, on sites with
different degree of damage in 1979. Per cent of trap
sites within the different groups of damage in 1980
compared to the degree of damage in 1979.
(Counties in south-east Norway).

Number of infested trees around the traps in 1979	New infested trees in 1980				
	None	1-5	6-30	More than 30	n
None	95	1	4	0	122
1-5	70	22	8	0	142
6-30	17	49	30	4	282
More than 30	26	10	40	24	143

within a short period after the infestation has taken place. This
is a major disadvantage of the trap-tree method and strengthens
the case for the use of pheromone-traps.

Another disadvantage of the trap-tree system is that the trap
-trees remove large numbers of parasites and predators of the bark
beetle. By selectively trapping bark beetles in pheromone baited
traps, the ratio of natural enemies to bark beetles is increased.
It is probable that this altered survivorship in turn may result
in higher bark beetle mortality rates.

Evaluation of the Method. Pheromone baited traps probably
will not supress outbreaks in overmature stands after an outbreak
has started. The only measure for such stands is clear-felling.
In areas with minor infestation, the traps may be useful in reducing the beetle-population to a level where they are too sparse to
overcome the resistance of living trees. This may have occurred
at some of the sites. New infestations in 1980 were recorded only
at 30% of the trap-sites which had minor damage in 1979.

At 70% of the sites with extensive damages in 1979, there
were new attacks also in 1980, despite the presence of traps, but
in most sites, the number of beetle-killed trees was reduced. Such
reductions are very important because the felling and removal of
newly infested trees in summer has been a bottle-neck in the control program.

No control measures have been undertaken over large areas,
particularly in steep terrain and where lack of roads makes felling and transport uneconomical. The total damage in cubic meters
in Norway in 1980 was about the same as in 1979. Reasons for this
may be that the outbreaks continue with full strength, particularly in inaccessible areas and in forests where the owner has not
been motivated to employ control measures.

The mass trapping program is not easy to evaluate from a
scientific viewpoint. We can estimate the number of trapped
beetles, but we are unable to determine the size of the natural
population. We can also compare tree mortality before and after
the trapping program, but we know little about the resistance of
the trees and the primary reason for mortality. The importance
of having control areas without traps is obvious. However, the
establishment of a research program which includes control areas
entails many difficulties, practical as well as scientific.

Pheromones will undoubtedly be a major component of most
future integrated bark beetle management programs. However, more
team-work is needed to get a better understanding on how the
method influences beetle populations.

Literature Cited

1. Bakke, A.; Frøyen, P.; Skattebøl, L. Naturwissenschaften 1977, 64, 98.
2. Vité, J.P.; Bakke, A.; Renwick, J.A.A. Can.Ent. 1972, 104, 1967-1975.
3. Krawielitzki, S.; Klimetzek, D.; Bakke, A.; Vité, J.P.; Mori, K. Z.ang.Ent. 1977, 83, 300-302.
4. Stoakley, J.T.; Bakke, A.; Renwick, J.A.A.; Vité, J.P. Z.ang. Ent. 1978, 86, 174-177.
5. Francke, W.; Sauerwein, P.; Vité, J.P.; Klimetzek, D. Naturwissenschaften 1980, 67, 147.
6. Bakke, A. Z.ang.Ent. 1981, 92(2). 172-177.
7. NOU 1979. "Granbarkbillen. (The spruce bark beetle)." Norges offentlige utredninger: Universitetsforlaget, Oslo, 1979(22); p 71.
8. Klimetzek, D. Allg.Forst- u. J.-Ztg. 1978, 149, 113-123.
9. Austarå, Ø. Meddr.Norsk inst. skogforsk. 1978, 34, 125-151.
10. Klimetzek, D.; Adlung, K.G. Allg.Forst- u. J.-Ztg. 1977, 148, 120-123.
11. Vité, J.P. Allg.Forstz. 1978, 33, 428-430.
12. Bakke, A.; Strand, L. Rapp. Nor.inst.skogforsk. 1981, 5/81, p 39.
13. Bakke, A.; Kvamme, T. J.Chem. Ecol. 1981, 7, 305-312.
14. Gmelin, J.F. "Abhandlung über die Wurmtrocknis", Verlag d. Crusiusschen Buchhandlung: Leipzig, 1778; p 176.

RECEIVED February 24, 1982.

Management of the Gypsy Moth with Its Sex Attractant Pheromone

JACK R. PLIMMER, B. A. LEONHARDT, and R. E. WEBB

U.S. Dept. of Agriculture, Beltsville Agricultural Research Center, Beltsville, MD 20705

C. P. SCHWALBE

Otis Methods Development Center, Otis Air Force Base, MA 02542

The sex attractant pheromone of the gypsy moth [Lymantria dispar (L.)] is disparlure (cis-7,8-epoxy-2-methyloctadecane). The natural attractant is the (+) enantiomer; it is a powerful attractant for male moths and is used throughout the United States as a bait in traps to detect infestations. A convenient and economic synthesis, recently reported, involves oxidation of an inactive unsaturated precursor with an optically active complex to yield an epoxide of high enantiomeric purity. The racemic form is as effective as the (+) enantiomer for disrupting mating. This technique for population suppression has been evaluated in experiments to compare the effects of formulation, dose rate and population density on its efficacy. In light infestations, gypsy moth mating is effectively suppressed. Microcapsules, laminated polymeric "flakes" and hollow fibers were compared as controlled-release formulations.

The gypsy moth [Lymantria dispar (L.)] has been continually extending its range throughout the northeastern United States and Canada since its accidental introduction in Massachusetts in 1869, and populations have frequently built up to defoliating levels. Areas of northern Maryland now lie within the major infested zone, and there are outlying infestations in Virginia and West Virginia, Ohio, Michigan, Wisconsin, Washington, California, and Nebraska. The zone extends as far as Ohio in the west. It is not known what factors may limit the ultimate extent of the infestation. Isolated infestations have also occurred in several western and midwestern states. In 1980, 5.1 million acres of forest were defoliated by the gypsy moth. In

1981, more than 12 million acres of trees from Maine to Maryland were stripped of their leaves by the larvae, or caterpillars. The cost of dealing with this insect by application of insecticides, by importation of parasites, and by survey traps has exceeded $100 million since its introduction from Europe.

The gypsy moth is a univoltine insect. Egg hatch occurs in late April or May. During the larval period, the insect feeds voraciously on about 500 species of vegetation although it is most commonly thought of as a pest of oak forests. The larvae pupate in June and emerge as adults in late June or July. The adults do not feed; their sole function is reproduction. The adult male actively searches for the adult female, who generally does not fly, but remains on the bark of a tree and emits a pheromone that attracts the male. The pheromone, disparlure, can be detected from a long distance by the male who has well-developed feathery antennae that function as olfactory organs. After mating, the females lay a cluster or mass of 100–800 eggs that hatch the following year.

The egg masses are buff colored and merge well with natural background. Transport of egg masses on recreational vehicles, plants, wood, or outdoor equipment may trigger new infestations, sometimes in remote sites. A network of traps is maintained throughout the country to detect the presence of the gypsy moth.

The gypsy moth affects large public and privately owned wooded areas and forests. Insecticide application is useful when the larvae appear, but its scale is limited by the cost of treatment. State and local authorities spend a great deal on such programs.

The application of pesticides is also limited by regulation, which imposes wide buffer zones, and by community action to veto pesticide application. Carbaryl (1-naphthyl N-methyl carbamate, Sevin), acephate (O,S-dimethyl acetylphosphoroamido-thioate, Orthene), diflubenzuron (N-[[(4-chlorophenyl)amino]-carbonyl]-2,6-difluorobenzamide, Dimilin), and other insecti-cides are effective against the larvae.

Additional control measures include use of Bacillus thuringiensis formulations and the release of parasites and predators. There is a continuing program to find new species that can be imported and released. A virus has also been registered for use against the insect. Work on release of sterile males is in an early stage.

There has been considerable progress in the utilization of the gypsy moth pheromone, disparlure (cis-7,8-epoxy-2-methyl-octadecane), as a component of a management program. The structure of the pheromone was announced in 1970 (1) and disparlure has been used since that time for trapping and mating disruption. This program was undertaken by the USDA and cooperating state universities and Departments of Agriculture, and I would like to focus on research undertaken primarily by the Agricultural Research Service, the Animal and Plant Health

Inspection Service of the USDA and the Maryland State Department of Agriculture during the last 4 or 5 years.

Pheromone Chemistry

The initial investigation of disparlure obtained from natural sources did not reveal its optical activity. However, subsequent bioassay of synthetic materials suggests that the naturally occurring material is (+)-disparlure (2-5).

Racemic material is available in bulk and is used for mating disruption in control programs. It costs about $400 per kilogram in large quantities. The (+) enantiomer is much more expensive, and a number of syntheses have been described that entail the use of optically active starting materials or intermediates (6-10). The superiority of the (+) enantiomer over racemic disparlure to attract males into traps has stimulated efforts to devise more convenient and economical syntheses.

A synthesis recently reported by Sharpless of MIT and his coworkers appears to be the most promising of those described since 1974 (10). Its advantage lies in the formation of an optically active epoxide in 90 to 95 percent enantiomeric excess. Thus, by avoiding separation of diastereomers, it eliminates a potentially difficult and wasteful feature of the synthesis.

Asymmetric epoxidation is accomplished by reaction of an allylic alcohol with tert-butyl hydroperoxide in the presence of D-(-) or L-(+)-diethyl tartrate and titanium tetraisopropoxide (Figure 1a). The orientation of the product can be predicted in advance; (-) tartrates epoxidize from the top face of the double bond when the bond is viewed in a horizontal plane and the carbinol group is on the right, as seen from the front of the plane. The (+) tartrates epoxidize from below the plane (Figure 1b).

The intermediate in this synthesis of disparlure, (Z)-2-tridecenol, is obtained by hydrogenation of 2-tridecynol in the presence of a palladium catalyst poisoned with barium sulfate and quinoline. Oxidation of (Z)-2-tridecenol with tert-butyl hydroperoxide/titanium tetraisopropoxide/D-(-)-diethyl tartrate gives (2R,3S)-epoxytridecanol in enantiomeric excess reported to be as much as 98 percent after recrystallization.

The synthesis is completed by oxidation of the alcohol to an aldehyde with a pyridine-chromic oxide complex. The reaction of the alcohol with a Wittig reagent, 4-methylpentylidene triphenylphosphorane, gives an olefin which can be hydrogenated to (+)-disparlure, (7R,8S)-epoxy-2-methyloctadecane.

The synthesis has been evaluated in a number of laboratories, including our own. Modifications may be made over several stages to improve yields but the basic features of the synthesis recommend it as a convenient and economical route.

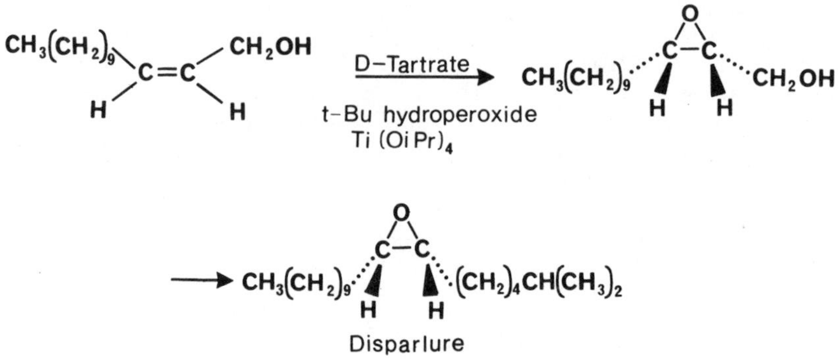

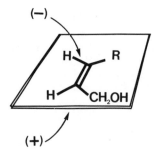

$$CH_3(CH_2)_9 \diagdown C=C \diagup CH_2OH \quad \xrightarrow[\substack{\text{t-Bu hydroperoxide} \\ \text{Ti (OiPr)}_4}]{\text{D-Tartrate}} \quad CH_3(CH_2)_9 \overset{O}{\overset{C-C}{\diagup \diagdown}} CH_2OH$$

$$\longrightarrow \quad CH_3(CH_2)_9 \overset{O}{\overset{C-C}{\diagup \diagdown}} (CH_2)_4CH(CH_3)_2$$

Disparlure

Figure 1a.   Asymmetric synthesis of (+)-disparlure.(Reproduced from J. Amer.
Chem. Soc.)

(−)

H   R

H   CH₂OH

(+)

Figure 1b.   Epoxidation in the presence of tartrates.

(+)-Disparlure is considerably more active than the racemic
material and is the most desirable attractant for trapping
males. Its use for this purpose is now widespread.

The Pheromone and Its Usefulness In Control Programs

Traps. The racemic form of the pheromone has been used for
a long time in survey traps placed throughout the United
States. The (+) enantiomer is now available in sufficient
quantity to provide a bait for many thousands of traps. Only
one-half a milligram is needed in each trap throughout a season
because disparlure is formulated in a long-lasting, controlled-
release device, usually a Hercon laminated polymeric dispenser.
A variety of traps containing (+)-disparlure as bait are
now appearing on the market. However, the value of trapping for
reducing populations requires much research before its value as
a control technique can be evaluated. This study is part of the
current ARS research program.

Mating Disruption. "Mating disruption" implies the use of
the pheromone to interfere with the reproductive cycle and
suppress a population by reducing the number of fertile eggs
laid by the females. In a pest management system, the method
could be used after the application of insecticide or Bacillus
to the larval stage of the insect. Mating disruption has the
advantage of not affecting natural predators or parasites of the
gypsy moth. Its disadvantage is that it appears to be more
effective at low population levels than in heavily infested
areas. The objective of our experiments was to discover the
most effective way of applying disparlure and to learn how its
effectiveness was affected by dose rate and population density.

Comparison of (+)- and Racemic Disparlure. (+)-Disparlure
was shown to be a better attractant than racemic disparlure
(4,5), and the addition of (-)-disparlure to (+)-disparlure
caused a decrease in trap catch, as is shown in Figure 2. In
our experiments in 1977, the trap catch with (+)-disparlure
increased as the amount of the (+) enantiomer used as lure was
increased, up to dose levels of 50-100 µg. Above this dose, the
catch declined with increased amounts of (+)-disparlure,
probably because of a disruptant effect.
We also wanted to know whether (+)-disparlure was more
effective than racemic disparlure as a mating disruptant. We
could not compare mating disruption by (+)- and racemic disparlure in a large-scale experiment such as was used to evaluate
formulations of disparlure because of the high cost of the (+)
enantiomer. However, we could obtain reproducible and statistically significant results for the reduction of male-moth catch
in small plots (25x25 m). Thirty-six cotton wicks suspended on
strings 2 m above the ground were treated with (+)- or racemic

disparlure (Figure 3). The reduction of catch in a central
array of 5 traps, each baited with (+)-disparlure, as a percentage of the catch in an untreated plot was used as a basis for
comparison.

Trap catches with 500, 5 and 0.05 mg racemic disparlure per
cotton wick showed reductions of 96, 84 and 38 percent respectively, while 5 mg (+)-disparlure gave 79 percent reduction.
Thus there was no significant difference in trap-catch
reductions obtained with 5 mg of racemic disparlure or of the
(+) enantiomer. Results are shown in Table I.

Formulation Development

Control of insects with pheromones depends on satisfactory
formulation. For the gypsy moth, a sprayable formulation was
needed that could be applied by air over rugged terrain.
Several years of research were spent developing formulations
that would release pheromone at a satisfactory rate through the
mating season (about 8 weeks).

Three types of controlled-release formulations were
ultimately used. These were:

(1) Microcapsules, a formulation based on plastic-
coated, gelatin-walled capsules (National Cash Register
Corp., Dayton, Ohio) encasing 3:1 xylene/amyl acetate
solution of 2.2 percent disparlure (NCR-2). The capsules
were suspended in water containing a thickening agent
(hydroxyethylcellulose), a sticker (1% RA-1645 , Monsanto
Corp., St. Louis, MO), and surfactant (0.1% Triton X-
202 , Rohm and Haas, Philadelphia, PA). Several similar
formulations were prepared: 1976-NCR-10 was based on
microcapsules containing 11% disparlure, and 1976-NCR-4 a
mixture of 3 parts of NCR-2 and 1 part of NCR-10. The
1978 NCR formulation was the same as 1976-NCR-2 with the
exception of the diameter of the microcapsules; the
former had a capsule diameter of 20-60 μ whereas all of
the 1976 formulations had capsule diameters of 50-250 μ
The formulations were applied using conventional aircraft
spray equipment.

(2) Hercon Disrupt , a formulation of polymeric 3-
layer laminated flakes ca 3 mm x 3 mm containing racemic
disparlure (20 mg/in^2) (Health-Chem Corp., New York,
NY). A sticker such as RA-1645 was incorporated in the
formulation which was applied aerially using specially
designed equipment.

(3) Albany International (Needham Heights, MA)
hollow fibers, a formulation of 11.5% by weight
disparlure in polyoxymethylene copolymer hollow fibers.
One kilogram of fibers was mixed with Bio Tac 3 as a
sticker. Specially designed equipment was used for
aerial application.

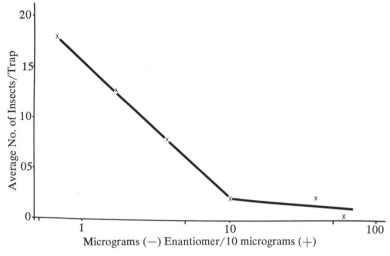

Figure 2. Decrease in trap catch caused by the addition of (−)-disparlure to the (+) enantiomer.

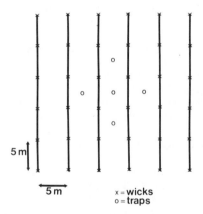

x = wicks
o = traps

Figure 3. Small plot design for tests to measure the reduction in male-moth catch. Key: wicks (×) and traps (○).

Table I. Male Gypsy Moth Trap Catch Reduction: Small Plot Tests*

Disparlure	Catch	%Reduction
500 mg. (±)	3.70 b	95.63
5 mg. (±)	13.73 b	83.79
0.05 mg. (±)	52.83 ab	37.65
5 mg. (+)	18.10 b	78.64
Control	84.74 a	▬

* Means followed by the same letter are not statistically different at the 95% level of confidence according to Duncan's multiple range test.

Many tests were conducted before reproducible results could be obtained, but by 1976, we were fairly confident that useful test protocols had been established. These have been improved, especially in relation to measurement of biological effects, and I would like to summarize our progress.

In 1976, we conducted tests to evaluate several formulations of disparlure for mating disruption. Each formulation was aerially applied to 4 replicate plots (16 ha each) at 20 g per ha in both Maryland and Massachusetts. Mating disruption was measured by the reduction both in the numbers of males caught in (+)-disparlure-baited traps in treated areas, as compared with untreated areas, and in the numbers of females mated in treated plots, compared with control plots. Ten laboratory-reared females were placed in each plot and were replaced approximately every 2 days. Recovered females were dissected to determine whether insemination had occurred. Eggs found at the site were examined after 4 to 6 weeks for evidence of embryonation. It was felt that we could not adequately demonstrate reduction or control of moth populations without further information relating to the growth and decline of insect populations; this was unobtainable without intensive population monitoring and a study of the quality of the population. Although control of moth populations might be achievable by a combination of pest management techniques, including pheromone application, our objectives in these tests were limited to comparison of the effects of disparlure treatments on the number of male moths caught in traps and the number of females mated.

This program formed the basis for tests in subsequent years. Greatly improved biological information is now available. Controlled-release formulations of disparlure have been improved. On the basis of this experience, we can now define where formulations of disparlure might play a role in gypsy moth control.

To improve the effectiveness of aerially applied formulations of racemic disparlure, further field tests were conducted in 1978 and 1979. In 1978, microencapsulated disparlure (1976 NCR) was applied at 50 g ai/ha to 16-ha plots, and a new formulation (1978 NCR) was applied at 5 g ai/ha and 50 g ai/ha.

The delta traps used gave only limited information because the efficiency of the trap falls off rapidly after 10 to 15 moths have been caught. The sticky surface soon becomes coated with insects and scales; subsequently many insects entering the trap may escape. Thus, catches tended to be lower than might be expected in untreated plots at peak flight. Improved traps of larger capacity were subsequently introduced.

The difference between the 1976 and 1978 NCR microcapsules used in the formulations lies in their diameter (50-250 µ for 1976 and 20-60 µ for 1978 formulations). Although the emission rate from the latter is higher, there was no difference in field performance at 50 g ai/ha.

There was a clear difference between the 5 g ai/ha and 50 g ai/ha treatments (1978 NCR). The latter caused a considerable decrease in mating and trap catch. The figures were 34 percent reduction in trap catch at 5 g ai/ha compared with 86 percent at 50 g ai/ha. Corresponding figures for mating reduction were 75.7 percent and 97.6 percent. A clear indication of response to dose was obtained.

In 1979, NCR microcapsules, Hercon flakes and Albany International hollow fiber formulations of racemic disparlure were applied by air in Massachusetts. Plots of 16 ha were treated with each formulation at 2 g ai/ha or 20 g ai/ha. Three plots served as controls, and each treatment was applied to three plots, except that during the application of the hollow fiber formulation, one plot received 10 g ai/ha instead of 20 g ai/ha. Data from this plot showed that trap catch and mating reduction fell between the values obtained for the 2 g ai/ha and 20 g ai/ha applications, providing a further indication that effectiveness increases with increasing application rates.

The reduction in trap capture for the NCR formulation at the 2 g ai/ha level was 44.6 percent. For the Hercon and Albany International formulations, it was 42.9 and 46.2 percent, respectively, at the same application rate. At 20 g ai/ha, the corresponding figures for the three formulations were 63.4, 75.7 and 91.6 percent (Table II). These figures were paralleled by the percentages of females that had been mated: at the 2 g ai/ha rate there was 33.6, 38.4 and 40.3 percent reduction in females mated. At the 20 g ai/ha rate, the corresponding figures were 62.6, 88.3 and 90.9 percent reduction for the three formulations. In the check plots, overall mating of females was 71.2 percent during the period of the experiment, and 1512 moths were caught in the traps. In the 1979 experiments, new large-capacity traps were used in place of delta traps; these did not suffer from the same defects as the smaller traps when large numbers of moths are present in the plots.

Evaluation of Tests for Mating Disruption

The role that pheromones will play in pest management programs will ultimately be determined by their efficacy, despite their many perceived advantages, which include their low potential for environmental pollution, the readiness with which they are degraded, their low toxicity to mammals, and the absence of harmful effects on beneficial insects. However, quantitation of efficacy as pest control agents has proved much more difficult with these behavior-modifying chemicals than it is with conventional insecticides because it must be based on measurements of their effects on population growth and decline that are not easy to obtain. The effects must be observed over a fairly large test area if statistically valid results are to be obtained.

Damage to forests by defoliation and tree kill is a clear indication of a heavy gypsy moth infestation, but since pheromone treatments are not effective at the high population densities present when such damage becomes apparent, reduction in damage cannot be used to determine efficacy of treatment. Egg mass surveys are another means used to determine levels of infestation, but at the low populations existing in suitable test areas they are not considered reliable as indicators of population density and thus cannot be used to obtain satisfactory criteria of effectiveness.

Evaluation of mating disruption tests was initially based on the reduction in numbers of male moths caught in disparlure-baited traps in treated areas, compared with the catch in areas that had not been treated. Within recent years, such traps have been baited with (+)-disparlure, a much stronger attractant. The frequency of mating of female moths provides the best measure of mating reduction. Virgin females reared in the laboratory and females that have emerged from field-collected pupae have been used for this purpose. In heavily infested areas, it may be possible to determine the relative numbers of females that are mated after emerging naturally in the test plots.

Measurement of larval population using burlap-band larval traps was found by Webb et al. (11) to provide a sensitive measure of relative gypsy moth populations in different plots. Older larvae feed in trees at night and leave the forest canopy to find convenient hiding places during the day. Thus, burlap bands on trees provide ideal sites for trapping. The older larvae within the plot tend to remain in the area, and since the adult females that subsequently emerge from pupae do not fly, the numbers of larvae trapped provide a good indication of the potential adult populations.

Reduction of male trap catch is a measure of the effect of pheromone treatment, but the efficacy of the treatment will be influenced by population density. Therefore, it was suggested that not only trap-catch reduction but also a measurement of population based on larval counts be taken into account to obtain an indication of the potential effect of pheromone treatment (11).

In Table III, mating communication disruption has been used as an indicator of the success of pheromone treatment. Male moth catch in (+)-disparlure-baited traps has been normalized against the number of larvae trapped pre-season, thus introducing a measure of relative population into the calculation of comparative effectiveness. The results obtained in 1980 in Maryland illustrate not only the variation of response with dose, but also the dependence of the efficacy of the technique on population levels.

Table II.   Reduction in Trap Capture and Mating Caused by Aerially Applied Racemic Disparlure

Gypsy moth:(±)disparlure treatment

7/22–8/20	No. of males caught	%Reduction in capture	%Females Mated	%Mating Reduction
Check	1512	▬	71.2	▬
NCR 2g/ha	837	44.6	47.3	33.6
Flake ll	863	42.9	43.8	38.4
Fiber ll	814	46.2	42.5	40.3
Fiber 10g/ha[1]	211	86.0	27.6	61.1
NCR 20g/ha	553	63.4	26.9	62.2
Flake ll	367	75.7	8.3	88.3
Fiber 20g/ha[2]	127	91.6	6.5	90.9

1   One plot at this dosage
2   Two plots at this dosage
Data from Schwalbe et al., 1981.

Table III.   Mating Communication Disruption* : Male Moth Catch Normalized Against Number of Larvae Found Under 50 Burlap Band Larval Traps (A Measure of Relative Population) Maryland 1980

Treatment (±)disparlure g/A	Larvae 50 traps	Male/plot 12 traps	Ratio Males/Larvae	%CD
30	471	222	0.94	92
12	1047	405	0.67	94
3	838	2531	4.96	59
▬	612	7905	12.08	

* %CD = [males/larva(control) − males/larva (treatment)]/[males/larva (control)] × 100

Conclusion

    The area defoliated by the gypsy moth increased
considerably during 1981. The use of disparlure in pest
management is primarily for detection of male moths. The (+)
enantiomer of disparlure is now available by a convenient
synthetic route and controlled-release formulations of (+)-
disparlure have replaced racemic disparlure as a bait in
traps. Small-plot tests suggest that the two forms would be
equivalent in their ability to disrupt mating. The pheromone
can be effectively used to reduce mating in lightly infested
areas. Our research in recent years suggests that it might be
worthwhile to evaluate higher dose levels of pheromone than
those that have been generally used. There are preliminary
indications that higher levels may be useful in more heavily
infested areas. This point of view has also been expressed by
Sower et al. (12) in their discussion of mating disruption
programs involving several forest insects.

Mention of a proprietary product is for information only and
does not constitute endorsement by the USDA.

Literature Cited

1. Bierl, B. A.; Beroza, M.; Collier, C. W. Science 1970,
    170, 87-9.
2. Yamada, M.; Saito, T.; Katagiri, K.; Iwaki, S.; Marumo,
    S. J. Insect Physiol. 1976, 22, 755-61.
3. Vité, J. P.; Klimetzek, D.; Loskant, G.; Hedden, R.; Mori,
    K. Naturwissenschaften 1976, 63, 582-3.
4. Miller, J. R.; Mori, K.; Roelofs, W. L. J. Insect Physiol.
    1977, 23, 1447-53.
5. Plimmer, J. R.; Schwalbe, C. P.; Paszek, E. C.; Bierl, B. A.;
    Webb, R. E.; Marumo, S.; Iwaki, S. Environ. Entomol.
    1977, 6, 518-22.
6. Iwaki, S.; Saito, T.; Yamada, M.; Katagiri, K. J. Amer.
    Chem. Soc. 1974, 96, 7842-4.
7. Mori, K.; Takigawa, T.; Matsui, M. Tetrahedron Lett. 1976,
    3853-6.
8. Farnum, D. G.; Veysoglu, T.; Cardé, A. M.; Duhl-Emswiler,
    B.; Pancoast, T. A.; Reitz, T. J.; Cardé, R. T.
    Tetrahedron Lett. 1977, 4009-12.
9. Pirkle, W. H.; Rinaldi, P. L. J. Org. Chem. 1979, 44,
    1025-8.
10. Rossiter, B. E.; Katsuki, T.; Sharpless, K. B. J. Amer.
    Chem. Soc. 1981, 103, 464-5.
11. Webb, R. E., and co-workers, unpublished.
12. Sower, L. L.; Daterman, G. E.; Sartwell, C. "Management of
    Insect Pests with Semiochemicals: Concepts and Practice",
    Mitchell, E. R., Ed., Plenum Press, New York, 1981, p. 351.

RECEIVED March 3, 1982.

Challenges in the Use of Pheromones for Managing Western Forest Lepidoptera

GARY E. DATERMAN, LONNE L. SOWER, and CHARLES SARTWELL

U.S. Dept. of Agriculture Forest Service, Pacific Northwest Forest & Range Experiment Station, Corvallis, OR 97331

Pheromones are rapidly taking their place as operational monitoring and suppression tools for managing western forest insect pests. Examples of operational uses include commercial formulations recently registered for suppression of western pine shoot borer, Eucosma sonomana, by mating disruption; and traps baited with a low-strength bait formulation for monitoring early build-up of Douglas-fir tussock moth, Orgyia pseudotsugata, populations. Remaining challenges for development of pheromones as suppression agents by the mating disruption technique include: population density effects, effects on the natural enemy complex, pheromone purity, limits on disruptant dosages, distribution of disruptant releasers, and a probable mode of action of the mating disruption technique. Discussions of releaser distribution and mode of action are based on recent field experiments with western pine shoot borer. Results indicate large, widely-spaced releasers are as effective as thousands of small releasers, and further suggest that false-trail-following is a likely factor contributing to the mating disruption effect. Discussion of monitoring populations with pheromone-baited traps emphasizes low-strength baits as a means of avoiding the trap-saturation problem, and describes how formulating baits may be complicated by the different release rates of compounds comprising multicomponent pheromones.

Two major objectives in preparing this paper were to provide an update of some recent successes in the development of pheromones for managing western forest Lepidoptera and to

discuss some of the remaining challenges related to field
applications of pheromones. The applications discussed will
include the mating disruption approach to suppress pest
populations by blocking their reproduction, and trapping
systems for monitoring populations.

Population Suppression

Probably the most exciting management application with
pheromones is the control of pests by interrupting or blocking
their normal reproductive behavior. In recent years there have
been several operational-level successes to control
lepidopteran pests by the mating disruption technique. This
approach involves the release of synthetic pheromone in some
type of controlled-release matrix. Normal male to female
chemical communicatiion is disrupted by the omnipresent
synthetic signals. A notable operational mating disruption
treatment has been the successful control of pink bollworm,
Pectinophora gossypiella, (Sanders), on cotton in the
southwestern United States (1, 2, 3). Attempts to control
western forest pests by mating disruption have been limited in
number but the experience with western pine shoot borer,
Eucosma sonomana, Kearfott, has been most encouraging. Two
companies, Albany International's Controlled Release Division
and Health Chemical's Hercon Division, have registered
formulations (mention of proprietary products do not
necessarily constitute endorsement by the USDA) which will
control this pest (4, 5, 6). Table I summarizes results of
controlling this insect with different formulations.

Table I. Efficacy of mating disruption for controlling western
          pine shoot borer in various field trials. (1980 data
          from personal communication with Dr. D. Overhulser,
          Weyerhaeuser Co., Centralia, Wash.).

Formulation	Year Treated	Hectares Treated	Pheromone (g/ha)	Damage Reduction (%)
PVC Strips	1978	8	3.5	83
Albany Int. Fibers	1978	20	15.0	67
Hercon Flakes	1979	100	20.0	88
Albany Int. Fibers	1979	600	10.0	76
Albany Int. Fibers	1980	634	5.0	75

We also hope to control Douglas-fir tussock moth, Orgyia
pseudotsugata (McDunnough), by mating disruption. Studies
dating from 1977 show that production of fertile eggs can be
greatly reduced by pheromone treatments (7). As illustrated in
Table II, population density of the pest apparently influences

the success of disruption treatments. Even at high densities, however, the mating disruption treatment successfully reduced fertile egg production by 77 percent. Thus, we are optimistic that this approach to suppression may also be useful for controlling populations of this insect.

Table II. Efficacy of mating disruption for low and high density Douglas-fir tussock moth populations. (Condensed from: Sower, L. L; Daterman, G. E.; Sartwell, C. <u>In</u> "Management of insect pests with semiochemicals," E. R. Mitchell, ed. Plenum Press, New York, 1981.)

	Disruption Treatment (g/ha pheromone)		
	0	9	36
	------------% egg reduction----------		
low-density population	0	92	100
high-density population	0	-	77

Many challenges remain for improvement and better understanding of the mating disruption technique for insect control. Here we discuss several points which should rank high on any list of further research needs for development of the technique.

Population Density. In general, mating disruption is more effective at lower pest densities. This is probably why the technique is so effective against western pine shoot borer, since typical populations of that species reach only several hundred adult moths per hectare. The population density factor is most significant, however, for pest species which fluctuate greatly in population numbers. For such pests it is important to know the mating disruption efficacies that can be expected at different population density thresholds. Both efficacies and population density thresholds will likely differ by species. Adult behaviors may differ, particularly with respect to the role of non-pheromone stimuli influencing mate-finding and courtship. It follows that those species most dependent on pheromone communication in their reproductive behavior, will be more susceptible to control by the mating disruption technique.

The Natural Enemies Complex. The high specificity of pheromones is often promoted as a major advantage for their use in control. There is generally assumed to be no adverse treatment effects on the complex of predators and parasites associated with the pest. Only scant data exist on this subject (8), however, and further research is needed. The

effect of disruption treatments on a pest's natural enemy
complex is important for reasons of environmental safety and
treatment efficacy. We do not want to endanger any beneficial
species by disruption treatment. A further consideration is
the influence a pest-specific pheromone treatment might have on
the relative effectiveness of the natural enemy complex.
Within limits, we believe control would be enhanced, because
the unharmed complement of natural enemies would now exert
their influence against a pest population diminished by the
disruption treatment. Following this logic, we can envision a
disruption treatment that resulted in a 90 percent reduction in
mating, ultimately being more efficacious than a non-specific
toxic pesticide that killed 98 percent of the pest insects plus
a large proportion of their key natural enemies. Research is
needed to quantify the effects of disruption treatments on
natural enemies of the pest. Better understanding of these
effects could ultimately lead to more efficient mating
disruption programs.

Pheromone Purity. An intriguing challenge concerns
chemical purity of disruptant formulations. For disruption
purposes, is it necessary to treat with the precise blend of
pheromone isomers and secondary components or is only the major
pheromone chemical necessary? The answer to this question
could be different for different pests, but in most cases it
could influence treatment costs. Our experience with the
western spruce budworm, Choristoneura occidentalis, Freeman,
suggests that only the major component of that pheromone is
necessary for disruption. Conversely, some recent work on an
orchard pest (9) indicates that a complete pheromone-blend
disruption treatment is more effective.

Disruptant Dosage. Another variable deserving of more
research consideration is the pheromone dosage applied for
mating disruption. We may have become overly-fascinated with
the virtue of treating with small quantities of pheromone.
Certainly the potency of pheromones is phenomenal, and it is
amazing to observe 5 grams of pheromone per hectare suppress
reproductive activity of a species for several weeks! On the
other hand, insect species differ in the quantities of
pheromone they produce and to which they have a sensitivity.
Consequently, we believe larger quantities - perhaps hundreds
of grams per hectare - should be considered and at least tested
for efficacy against certain pests. This consideration may be
especially appropriate for some forest pests, because the
resource to be protected may be considerably larger in quantity
of foliage than an agricultural crop. There could easily be
100-times the volume of foliage and air space requiring
disruptant coverage in a coniferous forest, for example, as
compared to the same acreage of lettuce or artichokes. This
raises economic questions, of course, but that should not
preclude the research. From the viewpoint of environmental

safety there should be no major problems associated with such applications. If lepidopteran pheromones are truly non-toxic, then the application of 200 grams (7 ounces) per hectare versus 5-20 grams should make little difference. Higher dosages should receive more attention by pheromone research and development specialists.

Distribution of Disruptant Releasers. A "best way" to distribute synthetic pheromone for mating disruption has yet to be described. The principal questions concern the number and relative strength of releasers needed per hectare. Because the scientific literature includes innumerable references to pheromone "air permeation" in disruption treatments, one might assume that maximizing numbers of uniformly distributed releasers would be preferred. Thus, we might envision releasers in the form of spray-sized particles causing a cloud of pheromone vapor over the treatment area. Logic would suggest that a maximum number of releasers would provide the most uniform and effective cloud.

For perspective, however, we must also consider the successful disruption treatments that have resulted from placement of relatively few releasers at wide spacings. Farkas et al. (10) reported disruption of cabbage looper, Trichoplusia ni (Hubner), with releasers spaced at up to 400 m. Hand applied releasers placed at intervals of 5-10 m and emitting ca. 3.5-14.0 g pheromone/ha/season have been successful in reducing damage caused by western pine shoot borer (5). Further, significant reductions in trap captures of western pine shoot borer were achieved with only one high-strength disruptant releaser per hectare (11). Therefore, air permeation by pheromone is not always required for disruption, because a single large releaser per hectare will obviously not provide a uniform cloud of pheromone vapor. Thus, there appear to be no firm guidelines for a "best-way" to apply disruption treatments. For a particular pest, the best application method will be dictated by benefit cost considerations, the pest's pheromone-related behavior, and characteristics of the resource to be protected. The challenge to the researcher is to determine what will work within each given set of circumstances.

Mode of Action. Better understanding of how the mating disruption technique disorients male moths and prevents their locating females would help in determining the most effective means for application of disruptant formulations. Results of some recent experiments with western pine shoot borer suggest that "false-trail-following" is a key factor in how the mating disruption technique works. That is, mate-seeking males orient to the strongest pheromone odor plume and follow it upwind to its source. Presumably when this source is a synthetic pheromone releaser, the male is exposed to a pheromone emission much stonger than that released by an unmated female. The male's sensory system becomes habituated to the synthetic odor

and he is consequently desensitized for detecting and orienting
to the lesser quantities of pheromone emitted by the female.

In a 1981 study on western pine shoot borer, three sets of
traps baited with synthetic pheromone in controlled release PVC
(polyvinyl chloride) pellets (12) were placed in ponderosa pine
plantations (16 or more ha in size), treated with (a) Health
Chemicals Hercon laminated flakes at some 20,000 flakes/ha with
a pheromone release capacity of 10 g/ha/season; (b) strips of
Hercon laminated plastic (30 x 2 cm) at 25 strips/ha stapled on
trees at 20 m intervals with a pheromone release capacity of
5 g/ha/season; and, (c) an untreated check plot. The Pherocon
II adhesive traps were baited with PVC pellets at five
different bait strengths of 0.001%, 0.01%, 0.1%, 1.0%, and 10%
pheromone by weight. Each of these baits was replicated seven
times in each disruption treatment for a total of 35 traps per
plantation and 105 overall. The traps were placed in lines of
five with each bait concentration represented once in each
line. The positions of the traps were randomized by bait
concentration within each line, and traps were spaced at least
20 m apart. All traps were left in place through 4-weeks of
the shoot borer's seasonal flight.

The results of this experiement are pertinent to the mode
of action question for two reasons. First, there was no
significant difference in disruption of male response between
plots treated by the smaller, more numerous flakes and the one
treated by the widely-spaced large releasers (Table III and
Figure 1). Secondly, and perhaps most significant, was the
capture of moths in the treated plots by traps baited with the
high strength pheromone pellets. The release rate of pheromone
from the 1 and 10% pheromone baits was apparently sufficient to
at least partially overcome the effects of the disruption
treatments. We estimate these bait concentrations released
pheromone at rates 100–1000 times higher than an average
unmated female. Pheromone release from the higher
concentration PVC bait pellets probably exceeded or rivaled the
release from the disruptant formulations. In particular, no
significant differences were found in captures by traps baited
with low-strength (.001 and .01%) PVC pellets in the untreated
check plot and traps baited with high-strength (10%) PVC
pellets in the two treated plots (Table III). Thus, by
increasing trap-bait strength as much as 10,000-fold,
equivalent numbers of males could be trapped in the disruption
plots!

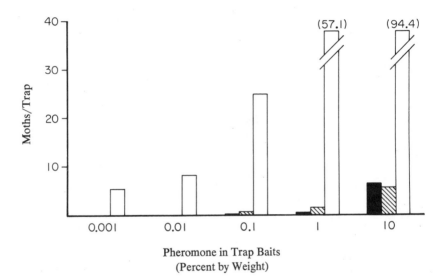

Figure 1.   Response of western pine shoot borer males to pheromone-baited traps in plots treated with different mating-disruption formulations and an untreated check plot. Key: check (□); 10 g pheromone in 20,000 flakes/ha (▨); and 5 g pheromone in 25 strips/ha (■).

We believe these results support a false-trail-following
theory for the mode of action of mating disruption. The fact
that males were able to discern and orient to a strong releaser
while in the center of areas treated for disruption,
demonstrates that omnipresence of pheromone vapor per se is not
always adequate to achieve mating disruption. Rather, the key
factor appears to be the relative strength of the pheromone
signals produced by trap-baits - or female moths - versus those
produced by disruptant particles.

Table III. Capture of western pine shoot borer males by traps
            containing pheromone-baits of different strengths
            and placed in plots treated for mating disruption.

Disruption Treatment	Pheromone Concentration in Trap-baits (%)	Average Males Captured/Trap[1]
Untreated Check Plot	10.0	94.4 (a)
	1.0	57.1 (b)
	0.1	25.0 (c)
	0.01	8.1 (d)
	0.001	5.3 (d)
10 g Pheromone in 20,000 Hercon Flakes/ha	10.0	5.7 (d)
	1.0	1.4 (e)
	0.1	0.6 (e)
	0.01	0 (e)
	0.001	0 (e)
5 g Pheromone in 25 Hercon Plastic Strips/ha	10.0	6.3 (d)
	1.0	0.4 (e)
	0.1	0.1 (e)
	0.01	0 (e)
	0.001	0 (e)

[1] Means followed by same letter not significantly
    different at .05 probability level. Evaluation by
    Analysis of Variance and Duncan's New Multiple Range Test.

These results suggest an upper limit might exist for the
number of particles used in formulations to release pheromone
in mating disruption applications. The greater the number of
particles in which a given dosage of pheromone is formulated,
the lower the release rate from any single particle (assuming
the duration of release remains constant). Disruption
applications might fail if release rates from disruptant
releasers fall below or near the release rate of female moths,
because the pheromone odor trails produced by females would

then rival those emitted from disruptant releasers. Thus, while coverage with synthetic pheromone releasers is important for successful mating disruption, we conclude that the relative strength of synthetic releasers versus the natural scent produced by the female is also a key consideration for development of formulations.

Population Monitoring

Trapping insects with pheromone-baited devices is now commonplace for many pest species. From a management viewpoint trap-captures can be very useful, simply by showing where populations exist and when they are active. More refined trapping systems may also be used to assess population densities and help predict where and when resource damage might occur. Special baits or traps may be necessary in the latter case, since the concern is with relative numbers caught and not just presence or absence of a species.

The Trap Saturation Problem. Traps with a limited capture capacity may become ineffective as monitoring devices when they fill with insects and can no longer reliably reflect continuing upward trends in population densities. One way to avoid this problem is by deliberately formulating a very weak trap-bait. Weak baits will attract few or no insects when low-density populations prevail, because of the bait's small active-space of attractiveness. Captures go up when pest numbers increase, simply because the increased numbers in flight increase the probability that some will intercept the active space of the monitoring traps. This approach to population monitoring has been successful as an early-warning system against approaching outbreaks of Douglas-fir tussock moth (Table IV).

Table IV.   Summary of Douglas-fir tussock moth population monitoring with pheromone-baited traps.

TUSSOCK MOTH POPULATION MONITORING

Year	No. Plots	Plots with > 20 Males/Trap	Follow-up "Sub-outbreak" Plots
1977	105	7	2
1978	118	9	7
1979	246	5	0
1980	550	3	2

The major value of this approach is that it signals an alert for only those areas where an outbreak might be developing. These areas are relatively few (Table IV), and consequently managers are able to focus their limited resources on these few locations.

There are, of course, other approaches to solving the trap-saturation problem for insect monitoring. Our main point here, however, is that the trap design used and the trapping objective may dictate the type of bait formulation and release rate of the attractant. It is not always best, for example, to formulate the strongest possible lure. For some survey or monitoring purposes, it may be necessary to match the female release-rate as nearly as possible, or, as in the case of the tussock moth, to formulate for a very low delivery rate of pheromone which will remain relatively constant over a long period of time. We believe these considerations represent a clear challenge to applications of pheromone-baited traps, particularly from the viewpoint of matching the release-rate of the trap-baits to their trapping objective.

Formulating Pheromone Blends

Lepidopteran pheromones usually consist of blends of two or more closely-related compounds. Investigations have become increasingly precise in identifying components of these blends and their proportions of different compounds and isomers. This precision, however, does not always carry over into the field bioassay or applications. Commonly, test compounds are loaded into bait releasers in the same ratios in which they are found being released by the insect. Release rates of these materials from the synthetic substrates, however, may vary considerably. Butler and McDonough (13) have quantified the release rates of a series of pheromone-related compounds from red rubber septa. Their results show marked differences in rates of release associated with differences in molecular chain length and oxygenated functional groups. A change of only two carbons in chain length may result in a 2 to 8-fold difference in release rate from rubber septa baits (13).

By holding uniform-sized PVC samples at constant temperatures and periodically weighing them, we were able to show differences in release rates of 14-carbon acetates, alcohols, and aldehydes. Table V summarizes these data, and shows that the aldehyde component is released about twice as fast as the acetate, with the alcohol being emitted at an intermediate rate.

Table V. Release rates of 14-carbon acetate, alcohol, and aldehyde pheromone components from a PVC matrix.

RELEASE RATES IN PVC PELLETS
(2% A.I., 22.5° C)

Compound	Average Release/cm/day[1] (ug)
E11-14:AC	8 (a)
E11-14:OH	10 (b)
E11-14:AL	15 (c)

[1] Means followed by same letter not significantly different at .05 probability level. Data evaluated by Analysis of Variance and Duncan's New Multipe Range Test.

Thus, release rates of at least some closely related pheromone components may vary greatly when formulated in either rubber septa or PVC pellets. If proper proportions of blend components are necessary to induce the appropriate behavioral response by a species, this influence of the formulation on release rates could be critical. We believe pheromone specialists need to be more alert to this potential problem, regardless of the type releaser substrate being used.

Literature Cited

1. Brooks, T. W.; Doane, C. C.; Staten, R. T. In "Odor Communication in Animals," F. J. Ritter, ed. Elsevier/North Holland Biomedical Press, Amsterdam 1979; pp. 375-388.
2. Brooks, T. W.; Doane, C. C.; Osborn, D. G.; Haworth, J. K. In Academic Press, Inc., New York, 1980; pp. 227-236.
3. Hennebery, T. J.; Gillespie, J. M.; Bariola, L. A.; Flint, H. M.: Lingren, P. D.; Kydonieus, A. F. J. Econ. Entomol. 1981, 74, 376-81.
4. Overhulser, D. L.; Daterman, G. E.; Sower, L. L.; Sartwell, C.; Koerber, T. W. Can. Entomo. 1980, 112, 163-5.
5. Sartwell, C.; Daterman, G. E.; Sower, L. L.; Overhulser, D. L.; Koerber, T. W. Can. Entomol. 1980, 112, 159-162.

6. Sower, L. L.; Overhulser, D. L.; Daterman, G. E.;
    Sartwell, C.; Laws, D. E.; Koerber, T. W. J. Econ.
    Entomol. (in press).

7. Sower, L. L.; Daterman, G. E.; Orchard, R. D.;
    Sartwell, C. J. Econ. Entomol. 1979, 72:739-742.

8. Sower, L. L.; Torgersen, T. R. Can. Entomol. 1979,
    111:751-2.

9. Charlton, R. E.; Carde, R. T. J. Chem. Ecol. 1981, 7,
    501-8.

10. Farkas, S. R.; Shorey, H. H.; Gaston, L. K. Environ.
    Entomol. 1974, 3, 876-7.

11. Daterman, G. E.; Sartwell, C.; Sower, L. L. In
    "Controlled Release of Bioactive Materials," R. Baker,
    ed. Academic Press, Inc., New York, 1980; p. 220.

12. Daterman, G. E. USDA For. Serv. Res. Pap. PNW-180, 12 p.,
    1974.

13. Butler, L. I.; McDonough, L. M. J. Chem. Ecol. 1981, 7,
    627-33.

RECEIVED February 24, 1982.

INDEX

INDEX

Jacket design by Kathleen Schaner.
Production by Karen Gray.

Elements typeset by Service Composition Co., Baltimore, MD.
Printed and bound by Maple Press Co., York, PA.